Electrical and Magnetic Properties
of Materials

ELECTRICAL AND MAGNETIC PROPERTIES OF MATERIALS

W Bolton

Longman Scientific & Technical

Longman Scientific & Technical,
Longman Group UK Limited,
Longman House, Burnt Mill, Harlow,
Essex CM20 2JE, England
and Associated Companies throughout the world.

First published 1992

British Library Cataloguing in Publication Data

Bolton, W.
  Electrical and magnetic properties of materials.
  I. Title
  620.1

  ISBN 0–582–07025–2

Set in Linotron Times 10/12 pt

Printed in Malaysia
by Percetakan Anda Sdn. Bhd.,
Sri Petaling, Kuala Lumpur

Contents

Preface

This book is an introduction to the electric and magnetic properties of materials, taking the reader from the basic concepts of atomic structure and electric and magnetic fields through a consideration of the mechanisms of dielectrics, magnetic materials, and the electrical properties of metals, semiconductors and insulators. The applications of such materials in electronics are discussed, such applications including the design of capacitors, liquid crystal displays, permanent magnets, transformer cores, photoconductive cells, the cathode ray tube, p–n junctions, injection lasers, light-emitting diodes, field effect transistors and other semiconductor devices.

The chapters in the book are arranged as follows:

Chapter 1   Overview of the basic elements of atomic theory necessary to appreciate many of the points in later chapters.

Chapters 2/3   Discussion of electric field principles and dielectrics.

Chapters 4/5   Discussion of magnetic field principles and magnetic materials.

Chapter 6   The motion of electrons in electric and magnetic fields.

Chapters 7/8   The movement of charge carriers in metals, semiconductors and insulators and across junctions.

Each chapter includes applications of the principles to electronic devices, worked examples and problems for the reader to tackle, answers being given to all.

The book is seen as being particularly relevant to students taking BTEC, HNC and HND courses in Electrical and Electronic Engineering, more than covering the sections K, L, M, N and O of the BTEC bank of Electrical and Electronic

Principles H and covering the associated parts of sections **B** and **I**. A knowledge of the basic principles of calculus and some dexterity with algebra has been assumed. Only a very basic knowledge of physical science and electrical principles has been assumed, the text generally developing ideas from basic principles, though the 'pace' may prove fast for those with only a very sketchy background.

W. Bolton

1 Atomic structure

Introduction

Why are some elements, e.g., copper and silver, good conductors of electricity and others such poor conductors? Why do some elements make good magnets, e.g., iron and cobalt, and others not? Why are some elements chemically very active, e.g., sodium and potassium (so active they must not be left in contact with air), and others inert, e.g., argon and neon (the so-called inert gases)? The answers to these and many other questions lie in a consideration of the atomic structure of elements. This chapter gives a brief overview of atomic structure which is adequate to begin to answer the above questions and also provide the basis for consideration of electric and magnetic properties of materials in later chapters.

The key to the understanding of the properties in terms of atomic structure is the description of such structures using quantum numbers and this is the approach adopted in this chapter. This may seem to be a mathematical trick, since it is difficult to imagine what the numbers are defining. However, the reality is that the behaviour of things on an atomic scale does not fit large-scale, easily imagined, models. Mathematics is the only way we can describe such behaviour.

Basic terms

A material that is made up of just one type of atom is called an *element*. Thus carbon is an element since it is made up of just carbon atoms. An *atom* is the smallest particle of an element that has the characteristics of that element. Atoms are themselves made up of other particles; these are however not characteristic of the element concerned but are basic building blocks from which all atoms are made.

Water is not an element since it is made up of hydrogen and oxygen atoms. The term *molecule* is used to describe groups of atoms, such as those forming water, which tend to exist together in a stable form. Some molecules are combinations of atoms from the same element. Thus, for example, gaseous hydrogen exists as molecules formed by two hydrogen atoms being combined together. Some molecules, like water, exist as combinations of atoms from a number of different elements. For example, the water molecule is two hydrogen atoms with an oxygen atom.

Atomic symbols

Since all the atoms of a particular element are the same type, a symbol can be used to represent one atom of that element. Thus, for example H is used to represent one atom of hydrogen, O represents one atom of oxygen, and C represents one atom of carbon. The letters are thus symbols for the elements but also represent fixed amounts of the elements, notably just one atom.

Molecules are formed from groups of atoms and, because a molecule always consists of a particular fixed combination of atoms, a chemical formula can be used to represent it. The chemical formula uses the symbols for the elements and indicates not only what elements are present but also the numbers of atoms of each that combine together to form the molecule. Thus the hydrogen molecule which consists of two hydrogen atoms is represented as H_2. The water molecule which consists of two hydrogen atoms with one oxygen atom is represented as H_2O.

Atomic structure

An atom consists of a nucleus surrounded by electrons. The *electrons* are negatively charged, each having a charge of -1.6×10^{-19} C. The nucleus contains *protons* which are positively charged, each carrying a charge of $+1.6 \times 10^{-19}$ C, and *neutrons* which have no charge. The nucleus has thus a positive charge as a result of the protons. The atom as a whole has no charge, since the numbers of electrons and protons in an atom are equal. Because unlike charged objects attract each other, attractive forces occur between the negatively charged electrons in the atom and the positively charged nucleus and these are the basic forces responsible for holding the particles together to form stable atoms.

The mass of each electron is 9.11×10^{-31} kg and this is much smaller than the masses of the proton and neutron. The mass of each proton and the mass of each neutron is about 1.67×10^{-27} kg, about 1830 times that of the electron. Thus most of the mass of an atom is contained within the nucleus. Because of this the number of protons and neutrons in the nucleus is a measure of the mass of an atom, hence the term *mass number* which is the sum of the numbers of protons and **neutrons** in the nucleus. Thus, for example, the hydrogen

Table 1.1 Basic constituents of some atoms

Element	Symbol	Atomic number	Mass number	electrons	Numbers of protons	neutrons
Hydrogen	H	1	1	1	1	0
			2	1	1	1
Carbon	C	6	12	6	6	6
			13	6	6	7
Oxygen	O	8	16	8	8	8
			17	8	8	9
			18	8	8	10
Silicon	Si	14	28	14	14	14
			29	14	14	15
			30	14	14	16
Copper	Cu	29	63	29	29	34
			65	29	29	36

atom has one proton and no neutrons, hence a mass number of one. Carbon has six protons and six neutrons, hence a mass number of twelve.

Elements are composed of atoms all having the same form. This means they all have the same number of protons in the nucleus. The *atomic number* of an element is the number of protons in each atom of that element. Thus, for example, hydrogen with just one proton has the atomic number of one. Carbon has six protons and so an atomic number of six.

Atoms of the same element that contain different numbers of neutrons are called *isotopes*. Thus, for example, carbon can form stable atoms with either six or seven neutrons and so has two stable isotopes. Table 1.1 shows the basic constituents of the atoms of some elements, only the stable isotopes being given.

Example 1

An isotope of germanium has an atomic number of 32 and atomic mass number of 74. How many electrons, neutrons and protons have the atoms of this isotope?

Answer

The atomic number is the number of protons, hence the number of protons must be 32. The number of electrons equals the number of protons, hence the number of electrons is 32. The atomic mass number equals the sum of the numbers of protons and neutrons, hence

74 = number of protons + number of neutrons
number of neutrons = 74 − 32 = 42

The Bohr model of atoms

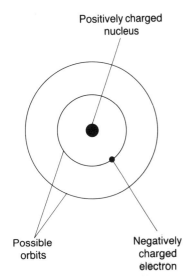

Positively charged nucleus

Possible orbits

Negatively charged electron

Fig. 1.1 The Bohr model of atoms

A simple model that has been used to describe the structure of atoms is the *Bohr model*. This considers atoms to consist of electrons moving in fixed orbits round a central nucleus, like planets orbiting the sun (Fig. 1.1). Only orbits at certain distances from the nucleus are possible and only when electrons are in these orbits is the arrangement stable. It is like saying that only at certain distances from the sun can we have planets in orbit. A planet at other distances would not be stable and would promptly move out of orbit.

The force holding an electron in orbit about a nucleus is the electric force of attraction between the negatively charged electron and the positively charged nucleus. The force of attraction between oppositely charged particles is inversely proportional to the square of the distance between their centres (Coulomb's law – see Ch. 2). This means that if the radius of the orbit is doubled then the force between the electron and nucleus is reduced to a quarter of its value. Thus electrons in orbits close to the nucleus have much stronger forces holding them in orbit than electrons in more distant orbits.

The different elements have different numbers of protons in the nucleus and different numbers of electrons. Hydrogen has just one proton and one electron and the Bohr model for this has the electron moving round the nucleus in the orbit which is closest to the nucleus (Fig. 1.2(*a*)). This innermost orbit is sometimes called the K shell. The K shell can hold just two electrons before it is full. Helium has two protons and two electrons and the Bohr model for this has the two electrons moving round the nucleus in the K shell (Fig. 1.2(*b*)). Lithium has three protons and three electrons and the Bohr model for this has two electrons in the K shell and one electron in the next possible orbit, the L shell (Fig. 1.2(*c*)). The L shell can hold eight electrons. Thus carbon with six protons and six electrons has two in the K shell and four in the L shell. Sodium has eleven protons and eleven electrons, hence the Bohr model has a full K shell with two electrons, a full L shell with eight electrons and one electron in the M shell (Fig. 1.2(*d*)). Table 1.2 shows the electronic structures for some atoms.

In arriving at his model Bohr used a so-called *quantum number n*. This number could have only the value 1, 2, 3, etc. and was the means by which he was able to specify that electrons could only exist in certain size orbits. When *n* equals 1 he arrived at the K shell, when *n* equals 2 the L shell, when *n* equals 3 the M shell. Bohr also assigned to each orbit a specific energy value for its electrons (see the discussion of energy levels later in this chapter).

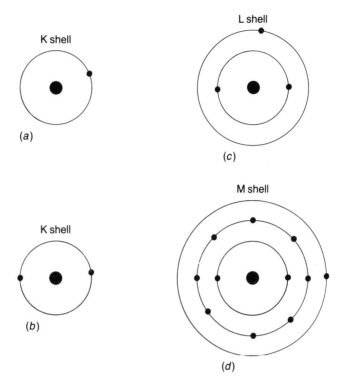

Fig. 1.2 The Bohr orbits for some atoms (a) Hydrogen, (b) Helium, (c) Lithium, (d) Sodium

Table 1.2 The electronic structure of atoms according to the Bohr model

Atomic number	Element	Electronic structure		
		K	L	M
1	Hydrogen	1		
2	Helium	2		
3	Lithium	2	1	
4	Beryllium	2	2	
5	Boron	2	3	
6	Carbon	2	4	
7	Nitrogen	2	5	
8	Oxygen	2	6	
9	Fluorine	2	7	
10	Neon	2	8	
11	Sodium	2	8	1
12	Magnesium	2	8	2
13	Aluminium	2	8	3
14	Silicon	2	8	4
15	Phosphorus	2	8	5
16	Sulphur	2	8	6
17	Chlorine	2	8	7
18	Argon	2	8	8

Quantum numbers

The above discussion of atomic structure has been in terms of the Bohr model. The Bohr model was only a partial success and has since been supplanted by a model which uses more quantum numbers. In the modern atomic model each electron has four *quantum numbers*. These numbers can only take certain values and thus what they are stating is that 'things' can only come in certain size packets, e.g., as with the Bohr atom only certain orbits are possible (Bohr's condition which led to only certain orbits being permitted was that angular momentum, i.e., the quantity mvr, where m is the mass of an electron, v its speed and r the orbit radius, can only have certain values). The quantum numbers are:

1 the *principal quantum number* n which can have any integer value of 1, 2, 3, 4, etc.;
2 the *angular quantum number* l which can have any integer value in the range 0 to $(n - 1)$, with an electron with $l = 0$ being called the s electron, $l = 1$ the p electron, $l = 2$ the d electron and $l = 3$ the f electron;
3 the *magnetic quantum number* m_l which can have the integer value, for each l value of 0, ±1, ±2, up to ±l;
4 the *spin quantum number* m_s which can have the value of +½ or −½.

The physical significance that we can imagine for these quantum numbers are that the principal quantum number n, as with the Bohr atom model, determines the orbit, the angular momentum quantum number l specifies the subdivisions of a particular n orbit, the magnetic quantum number m_l specifies the magnetic moment that is possible for the electron in its orbit (see Ch. 5), and the spin quantum number m_s specifies the magnetic moment possible for the electron due to its spin (see Ch. 5). There are only two possibilities for the spin quantum number, one indicating that the electron is spinning in one direction and the other that the electron is spinning in the exactly opposite direction (see Ch. 5).

Each electron in an atom has its own set of the four quantum numbers. No more than one electron in an atom can have a given set of the four quantum numbers, this statement being known as *Pauli's exclusion principle*. Thus if we start to build atoms up from $n = 1$, when all the possible combinations of quantum numbers have been used for that n value a new value of n has to be started.

The principal quantum numbers can be considered to specify the main shell with the K shell being $n = 1$, the L shell $n = 2$, the M shell $n = 3$, etc. Thus, for the K shell, a consequence of n being one is that since l has the value $(n - 1)$ it must be zero. The values of l can be considered to indicate the presence of subshells. Since there is only one value of l in this case then the K shell has all its electrons located in the same shell. Also, since l is zero, then m_l must be zero. The possible sets of quantum numbers when n equals one are thus:

1st electron $n = 1$ $l = 0$ $m_l = 0$ $m_s = +\frac{1}{2}$
2nd electron $n = 1$ $l = 0$ $m_l = 0$ $m_s = -\frac{1}{2}$

Only two sets of numbers are possible when n is one and so only two electrons can exist with this value of principal quantum number. In the Bohr model only two electrons could exist in the K shell. The spin quantum numbers indicate that one electron is spinning in one direction and the other in the exactly opposite direction.

Consider the L shell with $n = 2$. A consequence of this is that l can have the values 0 or 1, m_l the values 0, -1 or $+1$, and m_s $+\frac{1}{2}$ or $-\frac{1}{2}$. There are eight possible combinations of these numbers and hence eight electrons can exist with $n = 2$. The Bohr model has eight electrons in the L shell. However the quantum numbers lead to the concept of subshells, one subshell being for l with the value 0 (this being often called the s subshell) and the other for l with the value 1 (the p subshell).

	n	l	m_l	m_s	
1st electron	2	0	0	$+\frac{1}{2}$	2s subshell
2nd electron	2	0	0	$-\frac{1}{2}$	
3rd electron	2	1	0	$+\frac{1}{2}$	2p subshell
4th electron	2	1	0	$-\frac{1}{2}$	
5th electron	2	1	1	$+\frac{1}{2}$	
6th electron	2	1	1	$-\frac{1}{2}$	
7th electron	2	1	-1	$+\frac{1}{2}$	
8th electron	2	1	-1	$-\frac{1}{2}$	

For the M shell with $n = 3$ there are eighteen possible combinations of the four quantum numbers and hence eighteen electrons can exist in this shell. There are three subshells, 3s with $l = 0$ and consequently two electrons, 3p with $l = 1$ and six electrons, and 3d with $l = 2$ and ten electrons.

Table 1.3 shows the occupancy of the shells and subshells for elements. The electronic structure of atoms builds up in an orderly manner, from the K levels outwards, for the lower atomic number elements. However deviations occur from this, for example at potassium when the outer electron is in the N 4s shell rather than the M 3d shell.

Valence

The s subshells always require two electrons to fill them, the p subshell six, the d subshell ten. Elements with all their s and p subshells full are chemically inert in that they do not readily react with other elements. For example, the following are inert:

Table 1.3 The electronic structure of atoms

Atomic number	Element	K	L		M			N				O			P
	n =	1	2		3			4				5			6
		1s	2s	2p	3s	3p	3d	4s	4p	4d	4f	5s	5p	5d	6s
1	Hydrogen	1													
2	Helium	2													
3	Lithium	2	1												
4	Beryllium	2	2												
5	Boron	2	2	1											
6	Carbon	2	2	2											
7	Nitrogen	2	2	3											
8	Oxygen	2	2	4											
9	Fluorine	2	2	5											
10	Neon	2	2	6											
11	Sodium	2	2	6	1										
12	Magnesium	2	2	6	2										
13	Aluminium	2	2	6	2	1									
14	Silicon	2	2	6	2	2									
15	Phosphorus	2	2	6	2	3									
16	Sulphur	2	2	6	2	4									
17	Chlorine	2	2	6	2	5									
18	Argon	2	2	6	2	6									
19	Potassium	2	2	6	2	6		1							
20	Calcium	2	2	6	2	6		2							
21	Scandium	2	2	6	2	6	1	2							
22	Titanium	2	2	6	2	6	2	2							
23	Vanadium	2	2	6	2	6	3	2							
24	Chromium	2	2	6	2	6	5	1							
25	Manganese	2	2	6	2	6	5	2							
26	Iron	2	2	6	2	6	6	2							
27	Cobalt	2	2	6	2	6	7	2							
28	Nickel	2	2	6	2	6	8	2							
29	Copper	2	2	6	2	6	10	1							
30	Zinc	2	2	6	2	6	10	2							
31	Gallium	2	2	6	2	6	10	2	1						
32	Germanium	2	2	6	2	6	10	2	2						
33	Arsenic	2	2	6	2	6	10	2	3						
34	Selenium	2	2	6	2	6	10	2	4						
35	Bromine	2	2	6	2	6	10	2	5						
36	Krypton	2	2	6	2	6	10	2	6						
37	Rubidium	2	2	6	2	6	10	2	6			1			
38	Strontium	2	2	6	2	6	10	2	6			2			
39	Yttrium	2	2	6	2	6	10	2	6	1		2			
40	Zirconium	2	2	6	2	6	10	2	6	2		2			
41	Niobium	2	2	6	2	6	10	2	6	4		1			
42	Molybdenum	2	2	6	2	6	10	2	6	5		1			
43	Technetium	2	2	6	2	6	10	2	6	6		1			
44	Ruthenium	2	2	6	2	6	10	2	6	7		1			

Table 1.3 (cont'd)

Atomic number	Element	K			L		M				N				O			P
	n =	1			2		3				4				5			6
		1s	2s	2p	3s	3p	3d	4s	4p	4d	4f	5s	5p	5d	6s			
45	Rhodium	2	2	6	2	6	10	2	6	8		1						
46	Palladium	2	2	6	2	6	10	2	6	10								
47	Silver	2	2	6	2	6	10	2	6	10		1						
48	Cadmium	2	2	6	2	6	10	2	6	10		2						
49	Indium	2	2	6	2	6	10	2	6	10		2	1					
50	Tin	2	2	6	2	6	10	2	6	10		2	2					
51	Antimony	2	2	6	2	6	10	2	6	10		2	3					
52	Tellurium	2	2	6	2	6	10	2	6	10		2	4					
53	Iodine	2	2	6	2	6	10	2	6	10		2	5					
54	Xenon	2	2	6	2	6	10	2	6	10		2	6					
55	Caesium	2	2	6	2	6	10	2	6	10		2	6		1			
56	Barium	2	2	6	2	6	10	2	6	10		2	6		2			
57	Lanthanum	2	2	6	2	6	10	2	6	10	1	2	6		2			

Helium   1s full
Neon   1s, 2s and 2p all full
Argon   1s, 2s, 2p, 3s and 3p all full
Krypton 1s, 2s, 2p, 3s, 3p, 3d, 4s, and 4p all full

Elements which have just one electron in their outer s subshell are chemically very active. For example:

Lithium   1s full and one electron in 2s
Sodium   1s, 2s and 2p full, one electron in 3s
Potassium   1s, 2s, 2p, 3s, and 3p full, one electron in 4s
Rubidium   1s, 2s, 2p, 3s, 3p, 3d, 4s, 4p full, one electron in 5s

When one atom forms a chemical bond with another atom or atoms the aim is to achieve full outer s and p subshells and become inert. Thus a hydrogen atom with its single electron in the 1s shell is chemically active. It can become inert by combining with another hydrogen atom so that each atom shares electrons (Fig. 1.3) and so have complete shells. Sodium atoms have just a single electron in their outer 3s shell, while chlorine atoms have seven electrons (two in 3s and five in 3p) in their outer shell. Sodium chloride, common salt, is formed by a sodium atom donating its one electron to the chlorine atom (Fig. 1.4). This results in the sodium atom having full s and p shells and the chlorine atom having full s and p shells. The result is a stable compound.

Filled subshells are normally very stable arrangements in that atoms with just one electron outside a full subshell will readily give that electron to another atom to form a chemical

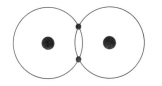

Fig. 1.3   The hydrogen molecule

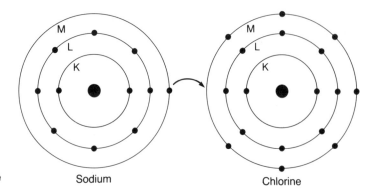

Fig. 1.4 Sodium chloride

Sodium

Chlorine

bond. Similarly, an atom lacking one electron to form a full subshell will readily accept an additional electron from another atom to form a chemical bond. Only the electrons in unfilled subshells take part in interactions with other atoms.

The *valence* of an element is equal to either the number of electrons in the outermost (s + p) shell of its atoms or the number of electrons needed to fill it. It is related to the ability of an atom to enter into chemical combinations with other atoms. Thus, for example, sodium having one electron in the s subshell of an otherwise empty M shell has a valence of 1. Carbon has two s and two p electrons in its L shell and so has a valence of four. Oxygen has two s and four p electrons in its L shell and because it needs just two further electrons to fill the p subshell has a valence of two.

Example 2

What is the valence of beryllium if its atoms have a full 1s shell and two electrons in 2s?

Answer

The outermost shell is the $n = 2$, or L shell, and it has a total of two electrons in that shell. The valence is thus 2.

Example 3

An element has the following arrangement of electrons in its atoms, what can be said about the activity of the element?

K shell: two electrons in 1s
L shell: two electrons in 2s
        six electrons in 2p
M shell: two electrons in 3s
        six electrons in 3p

Answer

The element has its outer s and p subshells full, two electrons being required to fill an s subshell and six for a p subshell, and so is inert.

The periodic table

The *periodic table* (Fig. 1.5) is an arrangement of the elements in the sequence of their atomic number so that each horizontal row of the table has elements with electrons in the same outer shell, i.e., having the same principal quantum number n or the same K, L, M, etc., designation (Fig. 1.6). Thus, for example, the first horizontal row (Table 1.4) consists of hydrogen, atomic number 1, and helium, atomic number 2. These both have atoms with their electrons in just one shell, the K shell (n = 1).

Table 1.4 The first horizontal row of the periodic table

	H	*He*
Atomic number	1	2
K subshell 1s	1	2
All other shells	0	0

The second horizontal row (Table 1.5) consists of the elements lithium, atomic number 3, to neon, atomic number 10. These all have atoms with full K (n = 1) shells and electrons in the L shell (n = 2).

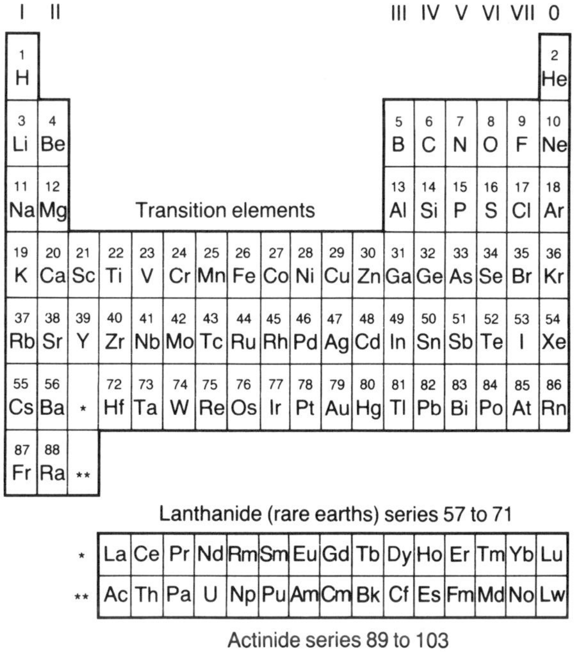

Fig. 1.5 The periodic table

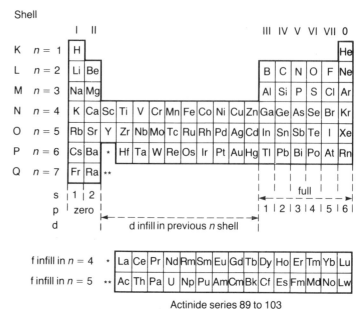

Fig. 1.6 Electron shell occupancy in the periodic table

Actinide series 89 to 103

Table 1.5 The second horizontal row of the periodic table

	Li	Be	B	C	N	O	F	Ne
Atomic number	3	4	5	6	7	8	9	10
K subshell 1s	2	2	2	2	2	2	2	2
L subshell 2s	1	2	2	2	2	2	2	2
L subshell 2p	0	0	1	2	3	4	5	6
All other shells	0	0	0	0	0	0	0	0

Elements with similar electronic structures for their atoms fall in the same vertical column (Fig. 1.6). The vertical columns are labelled as group I, II, III, IV, V, VI, VII, and 0 according to the number of electrons in the outermost shell. Group I elements have just a single electron in an s subshell (Table 1.6) while group II elements have full s shells with two electrons. Group III elements have full s shells and just one electron in the p subshell, group IV two electrons in the p subshell, group V three electrons, group VI four electrons, group VII five electrons and group 0 full p subshells with six electrons. This can be stated as: elements with the same valence are in the same vertical column (Fig. 1.7). Group I has a valence of 1, group II a valence 2, etc., with group 0 being the inert elements with zero valence since they have full shells.

A consequence of elements with similar electronic structures

Table 1.6 The electronic arrangements for group I elements

	K	L		M			N			O		P
	1s	2s	2p	3s	3p	3d	4s	4p		5s	5p	6s
Hydrogen	1											
Lithium	2	1										
Sodium	2	2	6	1								
Potassium	2	2	6	2	6		1					
Rubidium	2	2	6	2	6	10	2	6		1		
Caesium	2	2	6	2	6	10	2	6	10	2	6	1

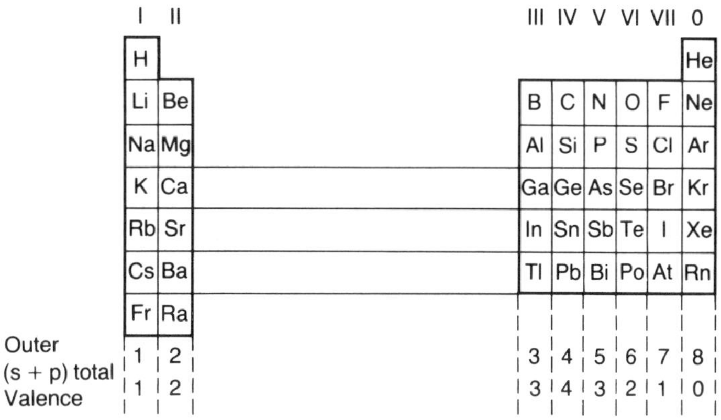

Fig. 1.7 Valence in the periodic table

being in the same vertical column is that all the elements in such a column have similar chemical properties. For example, all the very active elements are in column I and all the inert elements in column 0. Because all the elements in a vertical column have the same valence they form similar compounds. For example, the combinations of group I elements with chlorine to give chlorides results in a one-to-one relationship and so HCl, LiCl, NaCl, etc.

The term *transition elements* is used for those elements in the periodic table where inner d subshells are being filled in preference to the outer subshells. For example, scandium with atomic number 21 to zinc with atomic number 30 is one group of transition elements. For this group successive elements add electrons to the M 3d subshell though there are electrons in the N 4s subshell. All the transition elements can form more than one form of compound with other elements. For example copper can form with chlorine both CuCl and $CuCl_2$. Thus such elements have more than one valence.

The *lanthanides* or rare earths and the *actinides* are groups of elements where inner f subshells are being filled in preference to the outer subshells.

Example 4

On the basis of the periodic table what valence might be expected for germanium, atomic number 32?

Answer

Germanium is in group IV of the periodic table and so a valence of 4 would be expected.

Energy levels

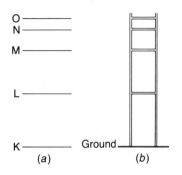

Fig. 1.8 The energy-level ladder for atomic electrons (*a*) the Bohr model, (*b*) the unequally spaced equivalent ladder

According to the Bohr model, electrons move in orbits round the nucleus. A moving electron will have kinetic energy, also an electron in such an orbit will have potential energy. Consider a negatively charged electron some distance from a positively charged nucleus: there is a force of attraction and so work must have been done to put the electron that distance from the nucleus. The situation is similar to the gravitational potential energy an object must have when lifted off the ground against the gravitational force of attraction and held some distance above the ground. Thus associated with an electron in an orbit is a particular amount of energy. Each orbit represents a particular *energy level*. Because only certain orbits are possible, an atom has only certain energy levels possible. We can think of the energy levels in the form of a ladder with electrons only able to exist on the rungs; the atomic ladder does not however have equally spaced rungs (Fig. 1.8).

This concept of an atom with energy levels in which its electrons can exist applies with the quantum number model of atoms. Figure 1.9 shows the energy level picture for subshells. For each value of *n* there is a set of levels corresponding to the s, d, p, f subshells. A point to be noticed is that there is overlapping of these sets of levels. For example, the 3d level is at a higher energy than the 4s level. Thus if the rule is that an electron would always end up on the lowest energy level where there is a vacancy, then the 4s level would be occupied before the 3d level. This is what occurs with the transition elements (see Table 1.3).

Thus the *n* quantum number can be thought of as defining a band of energies within which the *l* quantum number specifies the permissible levels. See Chapter 5 for an explanation of the other quantum numbers.

6p
5d
4f
6s

5p

4d
5s

4p

3d
4s

3p
3s

2p
2s

1s

Fig. 1.9 Energy levels for atomic subshells. This is not to scale but merely indicates the relative positions of the levels

Properties of elements

Why are some elements good conductors of electricity and others not? Copper, silver and gold are very good conductors of electricity. Table 1.7 shows the electronic arrangements in their atoms. They all show the same type of arrangements with a single electron in an outer s shell. Such an electron is relatively easily detached and so can give rise to an electrical current. This is a basic requirement for a good electrical conductor. Conversely, an insulator needs atoms which do not easily lose electrons. This is a simple picture of electrical conduction, see Chapter 7 for further discussion of this topic.

Table 1.7 The electronic arrangements of some good electrical conductors

Element	K	L		M			N				O			P
	1s	2s	2p	3s	3p	3d	4s	4p	4d	4f	5s	5p	5d	6s
Copper	2	2	6	2	6	10	1							
Silver	2	2	6	2	6	10	2	6	10		1			
Gold	2	2	6	2	6	10	2	6	10	14	2	6	10	1

Iron, cobalt and nickel, and some of their alloys and compounds, are widely used for permanent magnets. What are the characteristics of such elements that give them this property when other elements do not behave like this? Table 1.8 shows the arrangement of the electrons in the atoms of iron, cobalt and nickel. All three are transition elements with complete 4s shells and infilling, which is not complete, in the 3d shell. The distinguishing feature of all strongly magnetic materials is that they are transition elements, or in the case of compounds contain a transition element, and that such elements have the characteristic of an inner incomplete electron shell. See Chapter 6 for further discussion of this topic.

Table 1.8 Electronic configuration of permanent magnet elements

Element	K	L		M			N		
	1s	2s	2p	3s	3p	3d	4s	4p	4d
Iron	2	2	6	2	6	6	2		
Cobalt	2	2	6	2	6	7	2		
Nickel	2	2	6	2	6	8	2		

Example 5

On the basis of its arrangement of electrons in its atoms, would you expect sodium to be a good conductor of electricity?

Answer

Sodium has a full K shell, a full L shell and a lone electron in the s subshell of the M shell. Because of this lone s electron sodium would be expected to be a good electrical conductor.

Example 6

On the basis of its arrangement of electrons in its atoms, would you expect copper to be a strongly magnetic element?

Answer

Copper has the following arrangement of atomic electrons:

1s	2s	2p	3s	3p	3d	4s
2	2	6	2	6	10	1

Because it has no incomplete inner shells, copper would not be expected to be a strongly magnetic material.

Bonding between atoms

There are four ways by which atoms can bond together, these being ionic, covalent, metallic and Van der Waals'.

1 *Ionic bonding* is a result of electric forces of attraction between positive and negative ions, the ions occurring because of a transfer of one or more electrons from one atom to another.
2 *Covalent bonding* is a result of electrons being shared by atoms.
3 *Metallic bonding* is a result of atoms easily losing electrons and existing as positive ions in a sea of electrons.
4 *Van der Waals' bonding* is a result of atoms or molecules existing as either temporary or permanent dipoles.

Ionic bonding occurs with compounds such as sodium chloride (NaCl), which involve elements from groups I and VII combining, and magnesium oxide (MgO), which involve elements from groups II and VI combining. Ionic bonds involve an atom of a group I element donating an electron to an atom of a group VII element, as illustrated in Fig. 1.4. Because a group I atom has a solitary s electron, the result of its loss is that the atom has completely full shells. Also, because it has lost an electron the atom has then a net positive charge and is referred to as a positive ion. The atom of the group VII element has five electrons in its outer p subshell and thus gaining an electron results in the shell becoming full. After the gain the atom has a net negative charge and is referred to as a negative ion. There is then a force of attraction between the positive and negative ions and this is what constitutes the ionic bond. With the group II and VI elements, the atoms of the group II elements donate two electrons to those of the group VI elements to give full shells. The result is again positive and negative ions and an attractive force.

Atoms from elements in groups III, IV and V cannot easily transfer electrons since, because they have 3, 4 or 5 electrons in their outer shells, the energy required to move them is too great. However a different mechanism can be used for atoms to complete their outer shells – sharing electrons (see Fig. 1.3). The term *covalent bonding* is then used. Consider, for example, carbon. Carbon is in group IV and has:

K shell: full with two electrons in s
L shell: two electrons in s subshell and two in p subshell

To complete the p subshell it needs a further four electrons. This can be achieved if each carbon atom shares its L shell electrons with each of four neighbours. It shares an electron with each of the four neighbours and they, in turn, share an electron with it. Figure 1.10 shows a two-dimensional representation of this (see later in this chapter for further discussion).

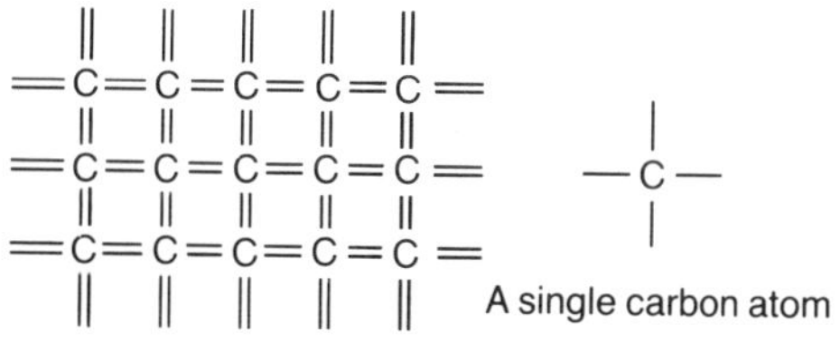

Fig. 1.10 Carbon atoms sharing outer shell electrons (each electron being represented by a dash)

Metals can be described as elements which have atoms which easily lose electrons to form positive ions. It is these freely available electrons which are responsible for the high electrical conductivity of metals. Such elements occur in the early groups of the periodic table. *Non-metals* are those elements in the later groups which tend to gain electrons to form negative ions and because they do not have easily detached electrons have poor electrical conductivities, being insulators. There are also some materials, called *near metals* or *semi-metals*, which are able to give reasonable electrical conductivity intermediate between that of the metals and non-metals. Metals occur in groups I, II and III, near metals in groups III and IV and non-metals in groups IV, V, VI, VII and 0 (Fig. 1.11).

All metals have electrons which are easily detached and so, for example, a piece of solid copper can be considered to be an array of positive ions existing in a sea of these detached electrons (Fig. 1.12). The positive ions are held together by their attraction to the surrounding electrons. This form of bonding is called *metallic bonding*.

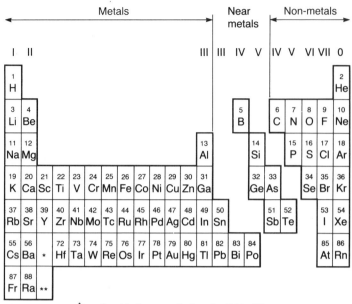

Fig. 1.11 Metals, near metals and non-metals in the periodic table

* Lanthanide (rare earths) series 57 to 71

** Actinide series 89 to 103

Fig. 1.12 Metallic bonding, the positive ions (+) being in a sea of electrons (−)

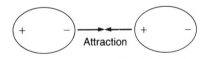

Fig. 1.13 Van der Waals' bonding

With inert gases and molecules, such as, for example, water, H_2O, there are no spare electrons available for any of the above types of bonding to occur. In the case of water there are no spare electrons because those of the atoms of hydrogen and oxygen are already involved in the bonds between the hydrogen and oxygen atoms. Yet solid water, ice, occurs and so there must be some form of bond between water molecules. The form of bonding that occurs between atoms and molecules having full shells is called *Van der Waals' bonding*. Such atoms can be thought of as essentially being a positive nucleus surrounded by a cloud of electrons. At any one instant the electrons may be uniformly distributed about the atom or more predominantly on one side. When this occurs the atom behaves as a temporary dipole. When two such atoms approach each other attraction can occur between these dipoles (Fig. 1.13). In the case of molecules, some are permanently dipoles because of the unequal distribution of charge (Fig. 1.14).

Example 7

Describe the form of the ionic bonds that occur with magnesium chloride ($MgCl_2$). The electronic arrangements of magnesium and chlorine atoms are:

	K		L		M	
	1s	2s	2p	3s	3p	
Magnesium	2	2	6	2		
Chlorine	2	2	6	2	5	

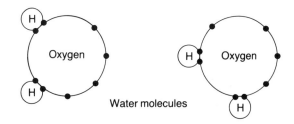

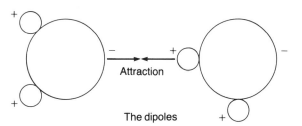

Fig. 1.14 Van der Waals' bonding between water molecules

Answer

Each magnesium atom gives up two electrons, those in 3s, to form a positive magnesium ion. Each chlorine atom gains one electron, completing its 3p shell, to become a negative ion. Two chlorine atoms are needed since there are two electrons given up by a magnesium atom. The result is magnesium chloride with two negative chlorine ions bonded to each positive magnesium ion.

Example 8

What type of bonds would you expect between (*a*) carbon and oxygen in the carbon dioxide molecule (CO_2) and (*b*) carbon dioxide molecules in solid carbon dioxide? The electronic arrangements of carbon and oxygen atoms are:

	K	L	
	1s	*2s*	*2p*
Carbon	2	2	2
Oxygen	2	2	4

Answer

(*a*) The bond is covalent with each carbon atom sharing in four electrons with each of the two oxygen atoms (Fig. 1.15).

(*b*) The bond is Van der Waals, since each carbon dioxide molecule has no spare electrons for any other form of bonding.

Example 9

Which of the following elements would you expect to form solids which are good conductors of electricity?

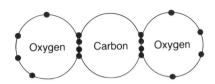

Fig. 1.15 The carbon dioxide molecule

The arrangement of atoms and molecules in solids

Barium (Ba), Iodine (I), Magnesium (Mg), Selenium (Se), Sulphur (S), Tungsten (W).

Answer

Barium, magnesium and tungsten are classified as metals and so good conductors of electricity. They form metallic bonds. Iodine, selenium and sulphur are classified as non-metals and so bad conductors of electricity. They form covalent bonds.

Solids in which atoms, or molecules, are arranged in an orderly manner are said to be *crystalline* and where there is an absence of order *amorphous*. The form of a crystal depends on how the atoms can be packed together in an orderly manner and this in turns depends on whether the bonds between the atoms are directional and the relative sizes of the atoms being packed together.

Ionic and metal bonds are formed between atoms without any directional inhibitions. We can consider such crystals to be essentially just orderly arrangement of spheres packed as close as possible. In the case of ionic crystals the spheres are the positive and negative ions and the crystal is held together by the electric forces of attraction between oppositely charged ions. In the case of metal crystals the spheres are positive metal ions held together by a 'glue' which is the sea of free electrons. Figure 1.16 shows ways by which spheres can be stacked to give the crystal forms known as *simple cubic*, *face-centred cubic*, *body-centred cubic* and *hexagonal close-packed*.

Covalent bonds are strong and directional. Carbon has a valence of 4 and the form of solid carbon known as diamond has each carbon atom bonding with four covalent bonds with neighbouring carbon atoms. Figure 1.17 shows the directions of such bonds and the resulting diamond structure. In general, the way the atoms are arranged in covalent bonded solids is determined by the directions of the bonds. Because the electrons are tied up in the bonds and are not free to move through the solid, covalent bonded solids are electrical insulators.

Solid carbon can however exist in a different form called graphite. Graphite, unlike diamond, is a good conductor of electricity. Figure 1.18 shows the way the carbon atoms are arranged in graphite. Of the four valence electrons of carbon atoms only three are involved in covalent bonds and so shared with other carbon atoms, in diamond all four are involved. The covalent bonds result in the carbon atoms forming into sheets of atoms which are relatively far apart. The unshared electron becomes unattached, like the free electrons in metals. The structure of graphite is thus of planes of covalent bonded atoms between which are free electrons. It is because of this that graphite is a good conductor of electricity.

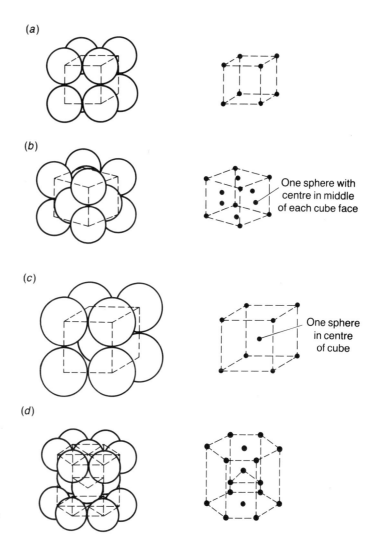

Fig. 1.16 Packing spheres in an orderly manner to give crystalline structures, (a) simple cubic, (b) face-centred cubic, (c) body-centred cubic, (d) hexagonal close-packed

One sphere with centre in middle of each cube face

One sphere in centre of cube

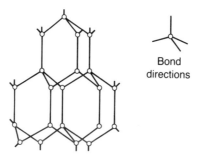

Bond directions

Fig. 1.17 Diamond structure

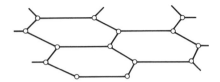

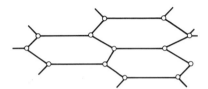

Fig. 1.18 Graphite structure

Example 10

Would you expect the electrical conductivity to be the same in all directions through the solid for (*a*) copper, and (*b*) graphite?

Answer

(*a*) Copper atoms are bonded together by just metallic bonds (face-centred cubic crystals) and the 'sea' of electrons in which the positive ions are embedded is not restricted to any one particular direction. Thus the conductivity might be expected to be the same in all directions since the electrons can move equally in all directions.

(*b*) Graphite has a sheet structure with the atoms in the planes of the sheets bonded together by covalent bonds. Between the sheets are free electrons. These free electrons are thus essentially constrained to movement parallel to the sheets and so the electrical conductivity in such a direction might be expected to be higher than in a direction at right-angles to the sheets. The conductivity is in fact about 30 000 times greater.

Polycrystalline or single crystal?

A piece of material which is just a single crystal has the same orderly packing, without break, through its entire structure. A polycrystalline material is made up of a number of crystals. Within each crystal the atoms are packed in an orderly manner but the boundaries between crystals present breaks in the orderly array so that the orderly array of one crystal changes in a random manner to the orderly array of the next crystal at such a boundary. Thus a polycrystalline material does not show the same continuous orderly packing throughout. Most solids are polycrystalline rather than single crystal, e.g., metals are polycrystalline. However, in some situations it is essential that the same orderly array persists throughout the material and thus a single crystal is required. This is the situation with silicon electronic devices (see Ch. 8).

Problems

1   What atoms and how many are present in the molecules (*a*) CO_2, (*b*) CH_4, (*c*) N_2?

2   Explain the terms atomic number, mass number and isotope.

3   State the numbers of electrons, protons and neutrons in the atoms of each of the following elements:
   (*a*) Calcium with atomic number 20 and atomic mass number 40
   (*b*) Iron with atomic number 26 and atomic mass number 56
   (*c*) Tin with atomic number 50 and atomic mass number 120.

4   Explain how the concept of shells, or orbits, for the electrons in atoms is used to explain the chemical stability of elements.

5   What values are possible for the quantum numbers and how do the combinations of these numbers lead to the idea of shells and subshells?

6   An element has two electrons in its outermost (s + p) atomic

shell, what will be its valence?

7 State the significance of the horizontal and vertical rows in the periodic table.

8 On the basis of the periodic table, what valence would be expected for (a) aluminium, (b) arsenic, (c) magnesium?

9 On the basis of the arrangement of the electrons in their atoms, would you expect (a) sulphur, (b) caesium, to be good conductors of electricity?

10 Explain what types of bonds might be expected for (a) aluminium atoms in the solid, (b) carbon atoms in diamond, (c) the atoms in potassium chloride (KCl), (d) the atoms in barium oxide (BaO), (e) sodium atoms in the solid, (f) solid helium.

11 Which of the following elements would you expect in the solid to be a good conductor of electricity? Chromium (Cr), Helium (He), Molybdenum (Mo), Antimony (Sb).

12 Electrical resistors are commonly made of carbon. Explain how the bonding between carbon atoms is responsible for carbon being a reasonably good conductor of electricity.

2 Electric fields

Introduction

Following some preliminary discussion of the charge concept, this chapter defines the electric field strength in terms of the potential gradient and from a consideration of the capacitance of a parallel plate capacitor leads to the concept of flux. An alternative approach would have been to start with a definition of electric field strength in terms of the force per unit charge, then perhaps a consideration of the force between point charges (Coulomb's law). In this chapter these evolve from the flux concept. This approach has been adopted because not only is it mathematically simpler but also puts the greater emphasis on the link between electric fields and potential gradients, a matter of some concern in later chapters.

Electric charge

If a plastic comb or pen, or other plastic object, is rubbed against a woollen object it becomes charged. We can tell it is charged because two such rubbed plastic objects exert repulsive forces on each other (Fig. 2.1). Like charged objects repel each other. Other objects when rubbed against other objects also become charged and they may be attracted to the rubbed plastic objects. Unlike charged objects attract each other. The convention is adopted of stating that the two types of charge are negative and positive charges.

Generally polymer materials, such as bakelite and nylon, become negatively charged when rubbed. The molecules of such materials are long chains held together by Van der Waals forces and have no loosely bound electrons. The molecular chains consist of carbon atoms to which are attached hydrogen atoms (Fig. 2.2). The covalent bonds between the hydrogen and carbon atoms mean that the electrons are concentrated between the carbon and hydrogen atoms and so the situation can be considered to be essentially a central core with a negative charge to which are attached positive hydrogen ions.

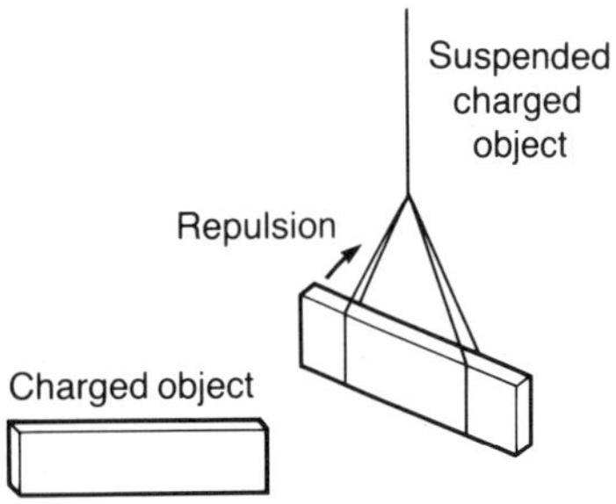

Fig. 2.1 Repulsive force between like charged objects

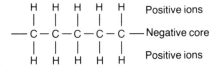

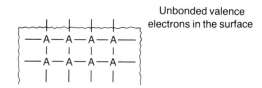

Fig. 2.2 Electrostatic charging with polymers involves the removal of hydrogen ions

The rubbing action removes hydrogen ions and so leaves a net negative charge to the polymer.

With other materials the mechanism is different. For example, with glass the rubbing leaves the glass with a positive charge. This can be explained by the surface atoms in glass losing electrons. With a solid made up of regular arrays of atoms the atoms at the surface will have been unable to complete bonds with neighbours and so have spare valence electrons (Fig. 2.3). It is these which become detached.

Fig. 2.3 Electrostatic charging with glass involves the removal of surface electrons

Polarisation

If a charged object, perhaps the rubbed plastic comb, is placed near small pieces of dry paper, or other insulators, attraction occurs (Fig. 2.4). This is despite the pieces of paper having no net charge. This can be explained by what is called *polarisation*. The molecules in the paper become distorted as a result of the presence of the charged object with the consequence that those parts nearer have opposite charge and those further away like charge. The molecules still have no net charge but because the opposite charges are nearer the attractive force predominates. The effect is sometimes called *induction*.

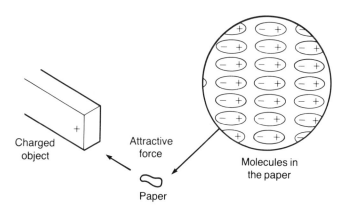

Fig. 2.4 Polarisation

Electric fields

A charged object repels or attracts other charged objects when there is some distance between them. In order to offer an explanation for this action at a distance a charged body is said to produce an electric field in the space around it. Any other charged body placed in the electric field experiences a force.

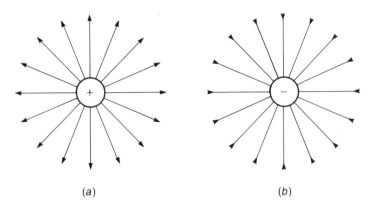

(a)　　　　　　　　　　　　　　(b)

Fig. 2.5 Electric field patterns, (a) an isolated positive charge, (b) an isolated negative charge, (c) adjacent charges of opposite polarity

(c)

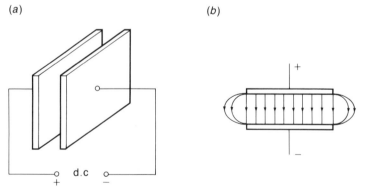

(a)　　　　　　　　　　　　　　(b)

Fig. 2.6 (a) Parallel conducting plates connected to a d.c. power supply, (b) the electric field pattern between the plates

An *electric field* is said to exist at a point if a charged object placed at that point experiences a force. The direction of the electric field at a point is the direction the force would be if a positive charge was placed at the point. Figure 2.5 shows some common field patterns. The lines so indicated are called *lines of electric force*. The tangent to any point of a line of force is the direction of the electric field at that point.

If two parallel conducting plates separated from each other by air are connected to the opposite terminals of a d.c. power supply (Fig. 2.6(*a*)) an electric field is produced between the plates. The pattern of this field is of straight lines at right-angles to the plates (Fig. 2.6(*b*)), except near the plate edges.

Electric field strength

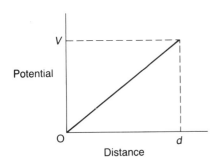

Fig. 2.7 Graph of potential with distance for the parallel plates

When a pair of parallel conducting plates is connected to a d.c. supply, as in Fig. 2.6, an electric field is produced between them. This can be detected by the force acting on a small charged object. With the exception of near the edges of the plates, the electric field strength is constant over that region between the plates. The electric field strength depends on the potential difference between the plates and their separation. However, a constant value of the potential difference divided by the distance gives a constant electric field. Measurements of the potential at points between the plates gives a graph of potential with distance in this electric field which is a straight-line graph (Fig. 2.7). The potential difference V between the plates divided by the distance d between the plates is the gradient of this graph.

Electric field strength is defined as the potential gradient at a point. Thus for the parallel plate arrangement when there is a potential difference of V between the plates and their separation is d, then the potential gradient is V/d. Hence

$$E = -\frac{V}{d}$$

The negative sign is because the direction of the electric field is in the opposite direction to that in which the potential is increasing. Consider the potential gradient to be rather like the gradient of a hill. When you are going up the hill and the potential is increasing, the electric field, like the gravitational field, is in the opposite direction. The unit of electric field strength is volt/metre.

In general terms, using calculus notation:

$$E = \text{potential gradient} = -\frac{dV}{dx} \qquad [1]$$

There will be a potential difference of dV between two points a distance dx apart in an electric field E, the electric field being assumed to be constant over this infinitesimally small distance. If we want to determine the potential difference between two points a distance x apart in an electric field, then we have to consider the distance as broken down into infinitesimally small segments, each of length dx, with the electric field being constant within a segment and then sum all the Edx terms in order to arrive at the potential difference. Thus potential difference $= - \Sigma$(all the Edx terms between $x = 0$ and x)

This can be written as

$$\text{potential difference} = - \int_0^x E\,dx \qquad [2]$$

Example 1

What is the electric field between a pair of conducting plates, 4 mm apart, when the potential difference between them is 200 V?

Answer

The electric field between the plates is the potential gradient, hence using equation [1]

$$E = - \frac{dV}{dx} = - \frac{V}{d} = - \frac{200}{0.004} = - 5.0 \times 10^4 \text{ V/m}$$

The minus sign indicates that the electric field is in the opposite direction to that of increasing potential.

Electric potential difference

The term potential energy is generally associated with mass and gravitational forces, an object having such energy by virtue of its position. The gravitational potential energy difference between two points is the work done in moving unit mass between those points, this work having to be done to move the mass against the action of the gravitational force acting on it. In order to return to its starting point the object would have to surrender this work. The term potential energy can however be used with electrical forces. In such a case the potential energy is the energy associated with the position of a charge. The electrical potential difference is defined in terms of the potential energy per unit charge. The *potential difference V* between two points is the work that has to be done to move a unit positive charge between them, this work having to be done to move the charge against the electric force that is acting on it.

$$V = \frac{\text{energy to move charge}}{\text{charge moved}} \qquad [3]$$

The unit of potential difference is the *volt* when the energy is in joules and the charge in coulombs. Thus if there is a potential difference of 10 V between two points in an electrical circuit then this means that 10 J are needed to move 1 C between them.

In the consideration of the movement of electrons, in particular within solids, a unit of energy called the *electron-volt* (eV) is often used. This is the energy acquired by an electron in being moved through a potential difference of 1 V. Since the charge carried by an electron is 1.6×10^{-19} C then

$$1 \text{ eV} = 1.6 \times 10^{-19} \text{ J}$$

If a charge is moved from a place where there is no electric force acting on it then we can consider such a place to be at zero potential and so the work done in moving a unit positive charge from there to some point gives the *electric potential* at that point. The place where there is no electric force acting on a charge has to be a place a long way from any other charged bodies and thus we talk of the electric potential at a point as being the work done in bringing a unit positive charge from infinity to the point.

Example 2

What is the work done in moving a charge of 0.1 C between two parallel conducting plates if there is a potential difference of 20 V between them?

Answer

Using equation [3]

$$\text{Work done} = VQ = 20 \times 0.1 = 2 \text{ J}$$

Electric force lines and equipotentials

A line or surface which passes through points which are all at the same potential is called an *equipotential*. Figure 2.8 shows the equipotentials, and electric lines of force, for a point isolated charge and a pair of parallel conducting plates which are connected to a d.c. supply. From the definition of potential, no work has to be done in moving an electric charge between two points at the same potential and so no work has to be done moving a charge along an equipotential line or surface. Thus there can be no electric force acting along an equipotential line or surface. This means that the electric force lines must be at right-angles to the equipotentials since only

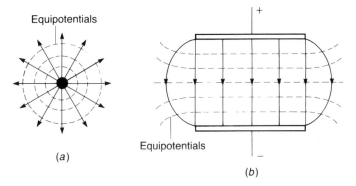

Fig. 2.8 Equipotentials for (a) a point isolated charge $+Q$, (b) parallel conducting plates

then will there be no electric force component along the equipotential line or surface.

An equipotential is a line of constant electrical potential energy per unit charge (see the earlier definition of potential difference) and so is like a contour line on a geographical map. Contour lines are lines of constant height and since gravitational potential energy is *mgh*, where *m* is the mass, *h* the height and *g* the acceleration due to gravity, then they are lines of constant potential energy per unit mass.

Capacitance

When a pair of parallel conducting plates, as in Fig. 2.6, are connected to a d.c. supply an electric field is produced between the plates. One of the plates becomes positively charged and the other negatively charged. The amount of charge on a plate depends on the potential difference V applied between the plates, Q being directly proportional to V for a given set-up of plates. Hence

$$Q = CV \qquad [4]$$

where C is the constant of proportionality. This constant is called *capacitance*. The unit of capacitance is the farad when V is in volts and Q in coulombs.

The factors determining the value of the capacitance for a pair of parallel conducting plates are the plate area A, the separation d of the two plates and the medium between the plates.

$$C = \frac{\varepsilon A}{d}$$

where ε is the factor, called the absolute permittivity, which relates to the medium used between the plates.

A more usual way of writing the equation is however in terms of how the permittivity of a material compares with that of a vacuum.

$$C = \varepsilon_r \varepsilon_0 \frac{A}{d} \qquad [5]$$

$$\varepsilon = \varepsilon_r \varepsilon_0 \qquad [6]$$

where ε_r and ε_0 are factors related to the medium between the plates. ε_0 is called the *permittivity of free space* and has a value of 8.85×10^{-12} F/m. ε_r is called the *relative permittivity*. It has no units, merely stating by how much ε_0 should be multiplied to account for the material between the plates not being a vacuum. For a vacuum ε_r has the value of one. For air the value is very close to one. For plastics the value is about two to three, for glass about five to ten. The insulating material used to separate charged surfaces is called a *dielectric*.

Typical values of relative permittivity are:

Dry air   1.0
Polythene   2.3
Dry paper   3.7
Mica   5.4

Example 3

What is the charge on the plates of a capacitor of capacitance 8 μF when there is a potential difference of 12 V across them?

Answer

Using equation [4], $Q = CV$, then

$$Q = 8 \times 10^{-6} \times 12 = 9.6 \times 10^{-5} \text{ C}$$

Example 4

What is the capacitance of a parallel plate capacitor which has plates of area 0.01 m^2, 5 mm apart with air as the dielectric?

Answer

The capacitance of a parallel plate capacitor is given by equation [5] as

$$C = \frac{\varepsilon_r \varepsilon_0 A}{d} = \frac{1 \times 8.85 \times 10^{-12} \times 0.01}{0.005} = 1.8 \times 10^{-11} \text{ F}$$

Dielectric strength

If the electric field strength in a dielectric is made high enough the dielectric will break down and become conducting. The maximum field strength that a dielectric can withstand is called the *dielectric strength* of the material. Because of this the electric field strength is sometimes called the *electric stress*. Typical values of dielectric strength are:

Air at normal pressure and temperature 3×10^6 V/m
Dry paper 1.6×10^7 V/m
Polythene 4×10^7 V/m

The field strength is the potential gradient. Thus for a given capacitor the greater the potential difference the bigger the electric field strength. The potential difference that can be applied across a capacitor must therefore be kept below some limit, this limit being when the electric field strength reaches the dielectric strength. See discharge from points later in this chapter.

Example 5

What is the maximum potential difference that can be applied to a parallel plate capacitor having a dielectric of thickness 2.0 mm if the dielectric strength is 4×10^7 V/m?

Answer

The dielectric strength is the maximum electric field possible. Thus since $E = V/d$ (equation [1]),

$$V_{max} = Ed = 4 \times 10^7 \times 0.002 = 8 \times 10^4 \text{ V}$$

Charge density

For parallel conducting plates of area A and a distance d apart, when a potential difference V is applied between them a charge $+Q$ is produced on one plate and $-Q$ on the other. For such an arrangement the capacitance C is given by

$$C = \frac{Q}{V} = \varepsilon_r \varepsilon_0 \frac{A}{d}$$

This equation can be rearranged to give

$$\frac{Q}{A} = \varepsilon_r \varepsilon_0 \frac{V}{d}$$

Q is the amount of charge spread over a plate area A, hence the quantity Q/A is sometimes called the *charge density* σ.

$$\sigma = \frac{Q}{A} \qquad\qquad [7]$$

Since the electric field strength E between the plates is given by V/d, then

$$\sigma = \varepsilon_r \varepsilon_0 E \qquad\qquad [8]$$

Example 6

What is the charge density on a pair of parallel capacitor plates 4 mm apart with air as the dielectric when the potential difference between them is 100 V?

Answer

The electric field strength E is the potential gradient and so the charge density is given by equation [8] as

$$\sigma = \varepsilon_r \varepsilon_0 E = \varepsilon_r \varepsilon_0 (V/d)$$
$$= 1 \times 8.85 \times 10^{-12} \times (100/0.004) = 2.2 \times 10^{-7} \text{ C/m}^2$$

Flux density

A useful way of considering a charge Q is as a source of something called *flux* which spreads outwards from it. The flux spreads out along the lines of force and for this reason they are sometimes called *streamlines*, i.e., lines along which the flux streams. A charge of 1 coulomb is considered to be a source of 1 coulomb of flux. Thus for a charge of Q coulombs there is a flux of Q coulombs.

The amount of flux passing at right-angles through an area A is called the *electric flux density D*. Thus for a pair of parallel conducting plates connected to a d.c. supply, as in Fig. 2.4, if the charge on a plate is Q then each plate can be considered as a source of flux Q. This flux then flows from the positively charged plate to the negatively charged plate along the streamlines. If a plate has an area A the electric flux density will be Q/A.

$$D = \frac{Q}{A} \qquad [9]$$

The unit of flux density is coulomb/metre^2.

If the plates are a distance d apart, then the capacitance C is given by (equations [4] and [5])

$$C = \frac{Q}{V} = \varepsilon_r \varepsilon_0 \frac{A}{d}$$

This can be rearranged to

$$\frac{Q}{A} = \varepsilon_r \varepsilon_0 \frac{V}{d}$$

But Q/A is the flux density D and V/d is the electric field strength E, hence

$$D = \varepsilon_r \varepsilon_0 E \qquad [10]$$

If there was a vacuum between the plates then ε_r would have the value one. Then the flux density would be $\varepsilon_0 E$. When the material has a relative permittivity ε_r then the flux density required for the same electric field strength is

$$\text{flux density in material} = \varepsilon_r (\varepsilon_0 E)$$
$$= \varepsilon_r \times \text{flux density in a vacuum}$$

$$\varepsilon_r = \frac{\text{flux density in material}}{\text{flux density in vacuum}} \qquad [11]$$

Electric field of a charged sphere

Consider a sphere of radius R with a charge Q. The total flux emitted by the sphere will be Q. We can consider the source of the flux to be effectively a point source at the centre of the sphere. At some distance r (where r is greater than the sphere radius R) from the centre of the sphere the flux will have

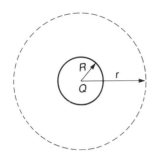

Fig. 2.9 Electric field at distance r from a charged sphere

spread out over an area of $4\pi r^2$ (Fig. 2.9). The flux density at radius r will thus be

$$D = \frac{Q}{4\pi r^2}$$

But $D = \varepsilon_r\varepsilon_0 E$ (equation [10]), where E is the electric field strength at radius r. Hence

$$\varepsilon_r\varepsilon_0 E = \frac{Q}{4\pi r^2}$$

$$E = \frac{Q}{4\pi\varepsilon_r\varepsilon_0 r^2} \qquad [12]$$

The electric field at radius r is independent of the radius of the charged sphere, provided r is greater than the sphere radius. The electric field at the surface of the sphere will however depend on its radius R since the bigger the value of R the greater the distance from the centre of the sphere at which the electric field is being considered. Thus

$$E \text{ at sphere surface} = \frac{Q}{4\pi\varepsilon_r\varepsilon_0 r^2} \qquad [13]$$

Example 7

Two spheres in the same medium have the same charge but one has twice the diameter of the other. How does the electric field strength at the surface of the double-diameter sphere compare with that at the surface of the other sphere?

Answer

The electric field strength at the surface of a sphere of radius R is given by equation [13] as

$$E = \frac{Q}{4\pi\varepsilon_r\varepsilon_0 r^2}$$

Since Q, ε_r and ε_0 are the same for both spheres

$$\frac{E_{2R}}{E_R} = \frac{R^2}{(2R)^2}$$

The electric field strength at the surface of the double-diameter sphere is one-quarter that at the surface of the other sphere.

Discharge from points

The electric field at the surface of a sphere is inversely proportional to the square of its radius. This means that for a given charge on the sphere the smaller the radius the greater the field strength at its surface. A pointed or sharp-edged object has a very small radius, the sharper it is the smaller the radius. Thus near a point the electric field will be much larger than near a flat or larger radius of curvature object carrying the same charge. A consequence of this is that the leakage of charge from pointed objects is much greater.

Ions are always present in air and in the presence of the electric field of a charged object these ions are acted on by forces and so accelerate, like charges moving away from the charged object and unlike charges to it. The consequential movement of the ions constitutes the leakage current. Thus a charged object placed in air loses charge through the air as a result of the movement of these ions under the action of the electric field from the charged object. The greater the electric field strength the greater the force acting on an ion and so the more it accelerates and the greater the leakage current. Thus because the electric field is higher near points than flat surfaces the leakage current is greater.

If the field is high enough the ions may acquire sufficient kinetic energy to be able to collide with air molecules and ionise them, i.e., knock electrons out of their atoms. These further ions may also be accelerated sufficiently to cause yet further ionisations. This then results in a massive increase in the leakage current and we refer to it as dielectric breakdown, the electric field responsible for this being called the dielectric strength (see earlier in this chapter). The higher field strength produced by points means that breakdown will more readily occur with points than flat or larger radius of curvature surfaces.

Electrostatic photocopier

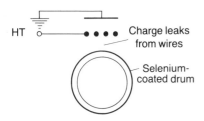

Fig. 2.10 Electrostatic photocopier

Figure 2.10 shows the basic form of the electrostatic photocopier. A high voltage is connected to a series of wires. Because these wires have small radii charge leaks from them to a selenium-coated drum which then becomes charged. The charged drum is then exposed to an optical image of the item being copied. Where light falls on the selenium it becomes conducting and allows the charge to leak away, where not illuminated it retains its charge. The result is an electrostatic charge pattern on the drum of the item being copied. This is then coated with a powder, the powder sticking as a result of electrostatic polarisation (see earlier in this chapter) to the charged areas. Thus when the drum presses against a sheet of paper a copy is produced.

Force on an isolated charge

Consider an isolated positive charge $+Q$ in an electric field and its move from an equipotential V to another equipotential $(V - \delta V)$ a distance δx away (Fig. 2.11). The potential gradient is $\delta V/\delta x$. This move to a decreasing potential means it

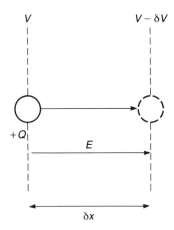

Fig. 2.11 Moving a charge in an electric field

is moving in the same direction as the electric field E.

$$E = -\frac{\delta V}{\delta x}$$

The charge will experience a force F as a consequence of being in the electric field. The work done is the force multiplied by the distance moved, hence

work done $= F\delta x$

The charge moves through a potential difference of $-\delta V$. Hence, since potential difference is the work done per unit charge, see earlier in this chapter and equation [3] for the definition of potential difference,

$$-\delta V = \frac{\text{work done}}{Q}$$

$$-\delta V = \frac{F\delta x}{Q}$$

$$E = -\frac{dV}{\delta x} = \frac{F}{Q}$$

Hence the force acting on a charge in an electric field is given by

$$F = EQ \qquad [14]$$

Example 8

What is the force acting on a small sphere carrying a charge of 3.2×10^{-19} C when it is in the air between the plates of a parallel plate capacitor which has a potential difference of 200 V applied between its plates? The plates of the capacitor are 4 mm apart.

Answer

The electric field E between the plates is the potential gradient V/d (equation [1]), hence the force is given by equation [14] as

$$F = EQ = (V/d)Q = (200/0.004) \times 3.2 \times 10^{-19} = 1.4 \times 10^{-14} \text{ N}$$

Coulomb's law

An isolated charge of $+Q_1$ in an electric field E will experience a force F, according to equation [14] where

$$F = EQ_1$$

Now consider that this electric field was produced by another isolated charge of $+Q_2$ (Fig. 2.12). The electric field at a distance r from such a charge is given by equation [12] as

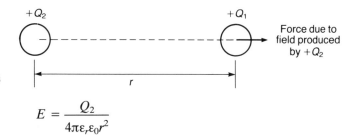

Fig. 2.12 Force between two charges

$$E = \frac{Q_2}{4\pi\varepsilon_r\varepsilon_0 r^2}$$

Hence

$$F = \frac{Q_1 Q_2}{4\pi\varepsilon_r\varepsilon_0 r^2} \hspace{3cm} [15]$$

This equation describes the force between two charges and is known as *Coulomb's law*.

Example 9

What are the forces acting on two spheres 200 mm apart in air when they are carrying charges of 2 µC and 3 µC?

Answer

Using Coulomb's law, equation [15],

$$F = \frac{Q_1 Q_2}{4\pi\varepsilon_r\varepsilon_0 r^2} = \frac{2 \times 10^{-6} \times 3 \times 10^{-6}}{4\pi \times 1 \times 8.85 \times 10^{-12} \times 0.200^2} = 1.3 \text{ N}$$

Electric field inside hollow conductors

The surface of a hollow conductor must be all at the same potential. If this were not the case and there was a potential difference between any two points then because the surface is a conductor a current would flow until the two points became the same potential. There is thus no potential gradient inside the hollow conductor, assuming there is no isolated charged object inside it. Because there is no potential gradient there is no electric field.

There is no electric field inside a hollow conductor due to any charge on its surface or to any external electric field. For this reason, a metal screen placed round an item will isolate it from the effects of an external electric field.

Capacitance of coaxial cylinders

A coaxial cable consists of a central circular conductor isolated from an outer tubular conductor by a dielectric, as in Fig. 2.13. It will have capacitance since it is essentially just two plates separated by a dielectric with a potential difference between them. The outer conductor shields the inside from the effects of any external electric fields. Hence the electric lines

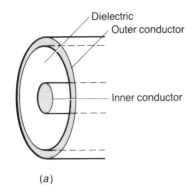

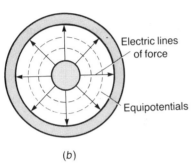

Fig. 2.13 (a) Coaxial cable, (b) electric field directions and equipotentials

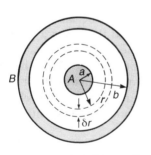

Fig. 2.14 Electric field at radius r

of force and the equipotentials are like those shown in Fig. 2.13(b), i.e., radial lines of force extending out from the inner conductor as far as the outer conductor and circular equipotentials in between the inner and outer conductors.

If the charge per unit length of the cable is Q coulombs/metre then the flux outflow from the inner conductor is Q coulombs/metre. At some radial distance r from the centre (see Fig. 2.14) this flux will have spread out over a cylindrical area of the circumference of the circle of radius r multiplied by the length of the cylinder. For a 1 m length then the area is $2\pi r$. Hence the flux density D at this distance is

$$D = \frac{Q}{2\pi r}$$

The electric field at this distance r is given by $D = \varepsilon_r \varepsilon_0 E$ (equation [10]), thus

$$E = \frac{D}{\varepsilon_r \varepsilon_0} = \frac{Q}{2\pi r \varepsilon_r \varepsilon_0}$$

The electric field strength is the potential gradient (equation [2]). This means that the potential difference δV across a small radial distance δr at radius r must be related to the field strength E by

$$E = -\frac{V}{\delta r}$$

Therefore

$$\delta V = -E\delta r = -\frac{Q}{2\pi r \varepsilon_r \varepsilon_0}\delta r$$

The total potential difference between the inner and outer conductors can be obtained by integration of this equation.

$$\int_B^A dV = -\int_b^a \frac{Q}{2\pi r \varepsilon_r \varepsilon_0} dr$$

$$V_A - V_B = -\frac{Q}{2\pi \varepsilon_r \varepsilon_0}(\ln a - \ln b)$$

$$V_A - V_B = \frac{Q}{2\pi \varepsilon_r \varepsilon_0} \ln \frac{b}{a}$$

The capacitance per unit length C of the coaxial cable is

$$C = \frac{Q}{V_A - V_B}$$

$$C = \frac{2\pi \varepsilon_r \varepsilon_0}{\ln (b/a)} \qquad [16]$$

The electric field strength at a radius r, where r is greater than a and less than b, is given by equation [12]

$$E = \frac{Q}{2\pi\varepsilon_r\varepsilon_0}$$

But $Q = CV$ and C is given by the expression above. Thus

$$E = \frac{V}{r \ln (b/a)} \qquad [17]$$

The electric field strength is inversely proportional to r, the smaller r is the bigger the electric field strength. The maximum value of the electric field strength will occur when r has its minimum value, i.e., $r = a$. Thus

$$E_{max} = \frac{V}{a \ln (b/a)} \qquad [18]$$

The minimum value of the electric field is when r has its maximum value, i.e., $r = b$. Thus

$$E_{min} = \frac{V}{b \ln (b/a)} \qquad [19]$$

Example 10

A coaxial cable has an inner core of radius 0.4 mm and an outer conductor of internal radius 5.0 mm. What is (*a*) the capacitance per metre of cable, (*b*) the maximum dielectric stress, if the dielectric between the inner and outer conductors has a relative permittivity of 2.6 and the potential difference between the two conductors is 10 kV?

Answer

(*a*) Using equation [16]

$$C = \frac{2\pi\varepsilon_r\varepsilon_0}{\ln (b/a)} = \frac{2\pi \times 1 \times 8.85 \times 10^{-12}}{\ln (5.0/0.4)} = 2.2 \times 10^{-11} \text{ F/m}$$

(*b*) The electric stress is the electric field intensity. The maximum value of this occurs when (equation [18])

$$E_{max} = \frac{V}{a \ln (b/a)} = \frac{10 \times 1000}{0.4 \times 10^{-3} \ln (5.0/0.4)} = 9.9 \times 10^6 \text{ V/m}$$

Capacitance of parallel transmission lines

Many transmission lines consisting of two parallel wires can be considered to be essentially two parallel cylindrical conductors separated by a dielectric (Fig. 2.15). There is a potential difference between them and when one has a charge per unit length of $+Q$ the other has $-Q$. Their distance apart D is much greater than the radii a of the wires. The electric lines of force and equipotentials between the two conductors will be

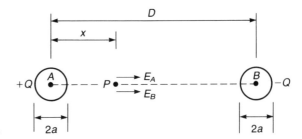

Fig. 2.15 Parallel transmission lines

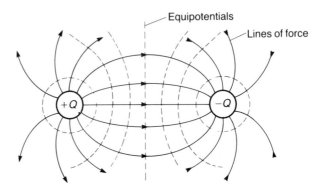

Fig. 2.16 Lines of force and equipotentials for parallel transmission lines

like that shown in Fig. 2.16, the equipotentials being at all points at right-angles to the lines of force.

The electric field strength E at some point P between the wires will be the sum of the electric fields due to each of the wires. For wire A the flux emitted per unit length is Q and this will have spread out over a cylindrical surface of area $2\pi x$ by P. Thus since the electric field strength is related to the flux density D by $D = \varepsilon_r \varepsilon_0 E$ (equation [10]) then the electric field at P due to wire A will be

$$E_A = \frac{D}{\varepsilon_r \varepsilon_0} = \frac{Q/2\pi x}{\varepsilon_r \varepsilon_0} = \frac{Q}{2\pi x \varepsilon_r \varepsilon_0}$$

Similarly, the electric field at P due to wire B will be

$$E_B = \frac{Q}{2\pi(D - x)\varepsilon_r \varepsilon_0}$$

Both these electric fields are in the same direction, hence

$$E = E_A + E_B$$

$$= \frac{Q}{2\pi x \varepsilon_r \varepsilon_0} + \frac{Q}{2\pi(D - x)\varepsilon_r \varepsilon_0}$$

$$= \frac{Q}{2\pi \varepsilon_r \varepsilon_0}\left(\frac{1}{x} + \frac{1}{D - x}\right)$$

But the electric field at point P is the potential gradient at that point. Thus a small change in distance of δx will mean a change in potential of δV, where

$$E = -\frac{\delta V}{\delta x}$$

The potential difference between the two wires, i.e., $V_A - V_B$, can be obtained by integrating this expression between the position when $x = a$ and when $x = (D - a)$.

$$\int_B^A dV = -\int_{D-a}^a E\, dx$$

$$\int_B^A dV = -\int_{D-a}^a \frac{Q}{2\pi\varepsilon_r\varepsilon_0}\left(\frac{1}{x} + \frac{1}{D-x}\right) dx$$

$$V_A - V_R = -\frac{Q}{2\pi\varepsilon_r\varepsilon_0}\left[\ln x - \ln(D-x)\right]_{D-a}^a$$

$$V_A - V_B = -\frac{Q}{2\pi\varepsilon_r\varepsilon_0}\{\ln a - \ln(D-a) - \ln(D-a) + \ln a\}$$

$$= \frac{Q}{\pi\varepsilon_r\varepsilon_0}\{\ln(D-a) - \ln a\}$$

If D is much greater than a then $\ln(D-a)$ approximates to $\ln D$ and hence $\ln(D-a) - \ln a$ becomes $\ln(D/a)$. Thus

$$V_A - V_B = \frac{Q}{\pi\varepsilon_r\varepsilon_0}\ln\left(\frac{D}{a}\right)$$

The capacitance per unit length C is

$$C = \frac{Q}{V_A - V_B} = \frac{\pi\varepsilon_r\varepsilon_0}{\ln(D/a)} \qquad [20]$$

Example 11

What is the capacitance per metre of two parallel wires in air, each wire being of diameter 4 mm and the distance between their centres is 40 mm?

Answer

Using equation [20] for the capacitance per metre

$$C = \frac{\pi\varepsilon_r\varepsilon_0}{\ln(D/a)}$$

Since the wires are in air ε_r can be taken as 1, thus with ε_0 being 8.85×10^{-12} F/m

$$C = \frac{\pi \times 1 \times 8.85 \times 10^{-12}}{\ln(40/4)} = 1.2 \times 10^{-11} \text{ F/m}$$

Capacitors in series and parallel

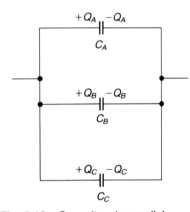

$$V_A \qquad V_B \qquad V_C$$
$$+Q \quad -Q \; +Q \quad -Q \; +Q \quad -Q$$
$$C_A \qquad C_B \qquad C_C$$

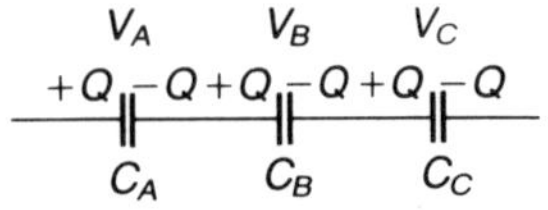

Fig. 2.17 Capacitors in series

$$+Q_A \quad -Q_A$$
$$C_A$$

$$+Q_B \quad -Q_B$$
$$C_B$$

$$+Q_C \quad -Q_C$$
$$C_C$$

Fig. 2.18 Capacitors in parallel

Consider three capacitors in series, as in Fig. 2.17. The potential difference V across the combination must be the sum of the potential differences across each capacitor.

$$V = V_A + V_B + V_C$$

The charge on one of the end plates of $+Q$ will induce a charge of $-Q$ on the opposing plate and also a charge of $+Q$ on the plate of the second capacitor which is connected to it and so on down the line. The result is that each capacitor will have plates with charges of $+Q$ and $-Q$. Dividing the above equation throughout by Q gives

$$\frac{V}{Q} = \frac{V_A}{Q} + \frac{V_B}{Q} + \frac{V_C}{Q}$$

But the capacitance of capacitor A is given by

$$C_A = \frac{Q}{V_A}$$

Similar expressions can be written for C_B and C_C. Since the overall capacitance of the system is

$$C = \frac{Q}{V}$$

then

$$\frac{1}{C} = \frac{1}{C_A} + \frac{1}{C_B} + \frac{1}{C_C} \qquad [21]$$

Now consider three capacitors in parallel, as in Fig. 2.18. The potential difference V across each capacitor will be the same. The charges on each capacitor will depend on their capacitances. The total charge Q is

$$Q = Q_A + Q_B + Q_C$$

Dividing each term by V gives

$$\frac{Q}{V} = \frac{Q_A}{V} + \frac{Q_B}{V} + \frac{Q_C}{V}$$

But for capacitor A

$$C_A = \frac{Q_A}{V}$$

Similar expressions can be written for B and C. Since the capacitance C of the system can be written as

$$C = \frac{Q}{V}$$

then

$$C = C_A + C_B + C_C \tag{22}$$

Example 12

What is the capacitance of systems consisting of (*a*) a 2 μF capacitor in series with a 4 μF, (*b*) a 2 μF in parallel with a 4 μF capacitor?

Answer

(*a*) For two capacitors in series (equation [21])

$$\frac{1}{C} = \frac{1}{C_A} + \frac{1}{C_B}$$

Hence

$$\frac{1}{C} = \frac{1}{2} + \frac{1}{4}$$

$$C = 1.3 \ \mu F$$

(*b*) For two capacitors in parallel (equation [22])

$$C = C_A + C_B$$

Hence

$$C = 2 + 4 = 6 \ \mu F$$

Multi-plate and multi-dielectric capacitors

Multi-plate capacitors are generally of the form:

plate–dielectric–plate–dielectric–plate–dielectric–plate

with alternate plates being connected together. In the above example there are four plates separated by three dielectrics to form three capacitors in parallel. In general, when there are *n* plates there are (*n* − 1) capacitors in parallel. Since they all have the same area *A* and dielectric thickness *d*, then the total capacitance is

$$C = \frac{(n - 1)\varepsilon_r\varepsilon_0 A}{d} \tag{23}$$

In some instances capacitors use dielectrics which are made up of layers of different materials, the form thus being

plate–dielectric 1–dielectric 2–plate

Each dielectric may be considered to form a capacitor and thus the arrangement is of a number of capacitors in series. Thus for the two-layer dielectric

$$C_1 = \varepsilon_{r1}\varepsilon_0 A/d_1$$

$$C_2 = \varepsilon_{r2}\varepsilon_0 A/d_2$$

where C_1 and C_2 are the two capacitances, ε_{r1} and ε_{r2} the respective relative permittivities and d_1 and d_2 their thicknesses. The charge on each capacitor will be the same. Thus

$$C_1 V_1 = C_2 V_2$$

where V_1 and V_2 are the potential differences across each dielectric. The field strength across each dielectric, E_1 and E_2, is the potential gradient, hence

$$E_1 = V_1/d_1$$

$$E_2 = V_2/d_2$$

Hence

$$\frac{E_1}{E_2} = \frac{V_1 d_2}{V_2 d_1} = \frac{C_2 d_2}{C_1 d_1} = \frac{\varepsilon_{r2}}{\varepsilon_{r1}} \qquad [24]$$

Curvilinear squares for estimating capacitance

Figure 2.19 shows the equipotential surfaces and electric lines cf force or streamlines for a parallel plate capacitor. At all points the equipotential surfaces and the streamlines are at right-angles. Also, since the conducting plates are also equipotentials the streamlines are, where they meet, at right-angles to them. Because of all these right-angles, tubes with

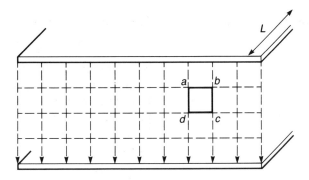

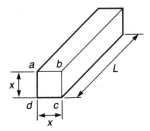

Fig. 2.19 Flux tubes for a parallel plate capacitor

rectangular cross-section ends are produced and by a choice of suitable spacing of the lines these can be made square cross-section tubes, as in Fig. 2.19. Such a tube is called a *flux tube*. The upper and lower surfaces of the tube are equipotentials, the sides streamlines. The entire space between the capacitor plates can be considered to be made up of square cross-section flux tubes.

With the parallel plates square cross-sections can be produced. However, in general the equipotentials and streamlines are curved. Thus, though the equipotentials and streamlines still meet at right-angles the sides of the square-like cross-section tubes are curved. These are referred to as *curvilinear squares*.

Consider the flux tubes shown in Fig. 2.19. If for the drawing of the equipotentials and streamlines a spacing of x has been chosen then each of the flux tubes will have a square cross-section with sides of length x. If the length of the flux tube, i.e., the length of the capacitor plates, is L then we can derive an equation for the capacitance C_t of the flux tube using equation [5] for the capacitance of a parallel plate capacitor. Equation [5] is really just the equation for the capacitance between equipotentials when the equipotentials happen to be those of the conducting plates. Hence

$$C = \frac{\varepsilon_r \varepsilon_0 \times \text{area of equipotential surface}}{\text{distance between the equipotentials}}$$

$$C_t = \frac{\varepsilon_r \varepsilon_0 L x}{x} = \varepsilon_r \varepsilon_0 L$$

Thus the capacitance of a flux tube with square cross-section is independent of the size of the square.

For the size of squares chosen in Fig. 2.19 the distance between the plates has been divided into four. This means that for each pair of streamlines there are four flux tubes in series. For four identical capacitors in series, each of capacitance C_t, the total capacitance C_s is

$$\frac{1}{C_s} = \frac{1}{C_t} + \frac{1}{C_t} + \frac{1}{C_t} + \frac{1}{C_t} = \frac{4}{C_t}$$

$$C_s = \frac{C_t}{4}$$

In general when the distance between the plates has been divided into n squares,

$$C_s = \frac{C_t}{n}$$

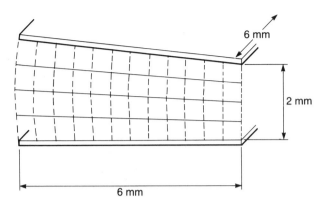

Fig. 2.20 Example 13

In Fig. 2.19 the width of the plates is such that there are ten squares across the width. This means that for each pair of equipotentials there are ten sets of flux tubes in parallel. Hence since the capacitance for capacitors in parallel is the sum of the capacitances, then for ten identical capacitors, each of capacitance C_s, the total capacitance C is $10C_s$ or $(10/4)C_t$. In general when the width is divided into m squares,

$$C = mC_s = \frac{mC_t}{n}$$

$$C = \frac{m\varepsilon_r\varepsilon_0 L}{n} \qquad [25]$$

For the capacitor shown in Fig. 2.19, if the plates have a width of 10 mm, a length of 12 mm and a separation of 4 mm then with squares of side 1 mm we have $n = 4$ and $m = 10$. Hence if the relative permittivity of the dielectric is 3.5 then

$$C = \frac{10}{4} \times 3.5 \times 8.85 \times 10^{-12} \times 0.012 = 8.9 \times 10^{-13} \text{ F}$$

This method can be used to estimate the capacitance of conductors with irregular configurations. In such situations there are likely to be approximations to one curvilinear square shape for the flux tubes.

Example 13

Estimate the capacitance for the non-parallel plate capacitor shown in Fig. 2.20 if the dielectric has a relative permeability of 3.0.

Answer

If squares of nominal side 0.5 mm are chosen then there will be twelve across the width, i.e., $m = 12$, and four across the distance between the plates, i.e., $n = 4$. The length of the flux tubes will be

6 mm. Hence, using equation [25],

$$C = \frac{m}{n} \varepsilon_r \varepsilon_0 \times L = \frac{12}{4} \times 3.0 \times 8.85 \times 10^{-12} \times 0.006$$

$$= 4.8 \times 10^{-13} \text{ F}$$

Energy stored in an electric field

Consider a capacitor being charged. Since $Q = CV$, then the charge Q on a capacitor plate is proportional to the potential difference V between the plates. Initially when there is no charge on the plates there is no potential difference and as the charge on the plates increases so the potential difference between the plates increases. When the potential difference between the plates is V then adding a small increment of charge δq means that work of $V\delta q$ has to be done (see earlier in this chapter for equation [3] and the definition of potential difference). But $V = q/C$ (equation [4]) so the work done for that increment of charge is $(q/C)\delta q$, where q is the charge on the capacitor at the instant when the increment of charge is added. The total work that has to be done in increasing the charge on the plates from 0 to Q is thus

$$\text{Energy} = \int_0^Q \frac{q\mathrm{d}q}{C} = \frac{Q^2}{2C} \qquad [26]$$

This energy is stored in the electric field and can be released when the capacitor is discharged.

The above equation can be written in a number of forms. Since $Q = CV$ then

$$\text{Energy} = \frac{Q^2}{2C} = \frac{(CV)^2}{2C} = \frac{CV^2}{2} \qquad [27]$$

Alternatively

$$\text{Energy} = \frac{Q^2}{2C} = \frac{Q^2}{2(Q/V)} = \frac{QV}{2} \qquad [28]$$

Another way of expressing this relationship is in terms of the electric flux density D. Since $D = Q/A$, where A is the plate area, then

$$\text{Energy} = \frac{QV}{2} = \frac{DAV}{2}$$

But the electric field E between the plates is given by $E = V/d$, where d is the separation between the plates. Hence

$$\text{Energy} = \frac{DAEd}{2}$$

But Ad is the volume of the space between the plates and so the volume of the electric field. Thus

$$\text{Energy/unit volume} = \frac{DE}{2} \qquad [29]$$

This can be written in terms of D or E since $D = \varepsilon_r \varepsilon_0 E$.

$$\text{Energy per unit volume} = \frac{D}{2}\left(\frac{D}{\varepsilon_r \varepsilon_0}\right) = \frac{D^2}{2\varepsilon_r \varepsilon_0} \qquad [30]$$

Alternatively

$$\text{Energy per unit volume} = \frac{\varepsilon_r \varepsilon_0 E \times E}{2} = \frac{\varepsilon_r \varepsilon_0 E^2}{2} \qquad [31]$$

Example 14

What is the energy stored in a capacitor of capacitance 2 μF when charged to a potential difference of 12 V?

Answer

The energy stored is given by equation [27] as
$$\text{Energy} = \tfrac{1}{2}CV^2 = \tfrac{1}{2} \times 2 \times 10^{-6} \times 12^2 = 1.4 \times 10^{-4} \text{ J}$$

Example 15

A capacitor with capacitance 4 μF is charged to a potential difference of 6 V. The dielectric between the plates of the capacitor has a cross-sectional area of 100 cm^2 and a relative permittivity of 2.5. What is (a) the electric flux density and (b) the energy stored per cubic metre of dielectric?

Answer

(a) Since $D = Q/A$ and $Q = CV$, then

$$D = \frac{CV}{A} = \frac{4 \times 10^{-6} \times 6}{100 \times 10^{-4}} = 2.4 \times 10^{-3} \text{ C/m}^2$$

(b) The energy stored per unit volume is given by equation [30] as

$$\text{energy/m}^3 = \frac{D^2}{2\varepsilon_r \varepsilon_0} = \frac{(2.4 \times 10^{-3})^2}{2 \times 2.5 \times 8.85 \times 10^{-12}} = 1.3 \times 10^5 \text{ J/m}^3$$

Problems

1 Sketch the electric field patterns, showing the lines of force and the equipotentials, for (a) a positively charged isolated conducting sphere, (b) a positively charged conducting sphere situated close to a similar negatively charged conducting sphere.

2 Define (a) electric field strength, (b) potential difference, (c) electric flux density, (d) capacitance.

3 A capacitor of capacitance 20 μF is charged to a potential difference of 6 V. What is the charge on the capacitor plates?

4 Two capacitors, with capacitances of 2 μF and 4 μF respectively, are connected in series across a 12 V supply. What will be (a) the charge on the plates of each capacitor and (b) the potential difference across each one?

5 Three capacitors, with capacitances of 2 μF, 4 μF and 8 μF respectively, are connected in parallel. What will be the total capacitance of the arrangement?

6 Derive, from the definitions of electric field strength as the potential gradient and potential, an expression for the force on an isolated charge in an electric field.

7 What is the size and direction of the electric field a distance in air of 40 mm from a small spherical conducting charge of $+ 2 \times 10^{-12}$ C?

8 State the factors which determine the capacitance of a parallel plate capacitor.

9 A capacitor consists of two square conducting plates of side 150 mm separated by a dielectric of thickness 2 mm. If the dielectric has a relative permittivity of 5 what is the capacitance?

10 A capacitor consists of two conducting plates, each having an area of 200 cm^2, separated by a dielectric of thickness 1.0 mm and having a relative permittivity of 3.7. What is (a) the capacitance and (b) the electric flux density when a potential difference of 300 V is applied between the plates, (c) the maximum potential difference that can be applied if the dielectric strength is 1.6×10^7 V/m?

11 What is the electric field produced between two parallel conducting plates, 4 mm apart, if a potential difference of 100 V is applied between them?

12 What is the electric field and electric flux density produced in the dielectric which fills the space between two parallel conducting plates if the plates have a separation of 2 mm, the dielectric has a relative permittivity of 2.5 and the potential difference between the plates is 100 V?

13 Sketch the electric lines of force and equipotentials for a coaxial cable when there is a potential difference between the inner and outer conductors.

14 A coaxial cable has an inner conductor of radius 2.0 mm and an outer conductor of internal radius 4.0 mm. If the dielectric between the conductors has a relative permittivity of 2.5 what is (a) the capacitance of a 3.0 m length of the cable, (b) the maximum and minimum values of the electric stress with a potential difference of 1 kV?

15 Two parallel wires in air have a separation between their centres of 50 mm. What is the capacitance of 200 m if the wires have a radius of 2.0 mm?

16 Explain how the method of curvilinear squares can be used to estimate the capacitance of irregular conductors.

17 What is the energy stored in a capacitor of capacitance 2 μF when it is charged to 10 V?

18 What potential difference must be applied across a 500 pF capacitor if it is required to store 0.5 J of energy?

19 A parallel plate capacitor has plates of area 3.0 mm apart in air.

What will be (*a*) the electric flux density and (*b*) the energy stored per cubic metre in the dielectric when a potential difference of 1.0 kV exists between the plates?

3 Dielectrics

Introduction

This chapter is about dielectrics, such materials being electrical insulators with relative permittivities greater than one. In order to obtain an understanding of the behaviour of such materials it is necessary to consider the atoms and molecules in such materials and their arrangement to form electric dipoles. Capacitors, piezo-electricity, crystal-controlled oscillators, ferroelectricity and liquid crystals all follow from a consideration of such principles.

Effect of a dielectric

Fig. 3.1 A dipole

The term *dipole* is used for atoms or groups of atoms that effectively have a positive charge and a negative charge separated by a distance (Fig. 3.1). Dipoles may be permanent because of an uneven distribution of charge in a molecule or induced as a consequence of an electric field being applied (see polarisation, Ch. 2). Whatever the origin of the dipoles, when an electric field is applied they will endeavour to line up with the field direction (Fig. 3.2). It is the same effect as a magnetic compass needle lining up with the direction of a magnetic

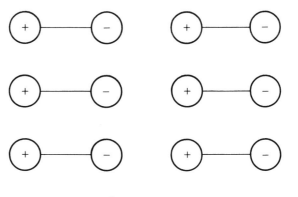

Fig. 3.2 Dipoles in an electric field

Electric field

field. When the dipoles line up the material is said to be *polarised*.

When a potential difference is applied across the plates of a parallel plate capacitor with a vacuum between the plates, charge accumulates on the plates. The amount of charge Q_0 is determined by the potential difference V since

$$Q_0 = C_0 V$$

where C_0 is the capacitance with a vacuum. However if there is a dielectric between the plates (Fig. 3.3) the alignment of the dipoles in it means that some of the charge on the plates is cancelled by the charge of the dipoles adjacent to the plates. This smaller charge Q on the plates for the same potential difference V means that the presence of the dielectric has increased the capacitance to C.

$$Q = CV$$

The factor by which the capacitance is increased is called the relative permittivity ε_r.

$$C = \varepsilon_r C_0 \tag{1}$$

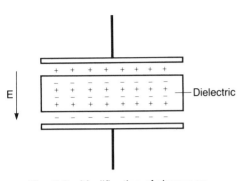

Fig. 3.3 Modification of charge on plates due to the presence of a dielectric

Polarisation

With a vacuum between the plates of a parallel plate capacitor the flux density D between them when there is an electric field E is given by (see Ch. 2, equation [10]):

$$D = \varepsilon_0 E$$

where ε_0 is the permittivity of free space, having the value of 8.85×10^{-12} F/m. If however there is a dielectric between the plates then for the same electric field the flux density is increased as a consequence of dipoles aligning themselves with the field. We can represent this by adding a term P to the above equation.

$$D = \varepsilon_0 E + P \tag{2}$$

P is called the *polarisation* and has units of C/m^2. It represents the combined effects of a considerable number of small dipoles aligning with the electric field. Since with a dielectric we can also write (Ch. 2, equation [10]),

$$D = \varepsilon_r \varepsilon_0 E$$

where ε_r is the relative permittivity of the dielectric, then

$$\varepsilon_r \varepsilon_0 E = \varepsilon_0 E + P$$

$$P = \varepsilon_0 E(\varepsilon_r - 1) \tag{3}$$

Example 1

What is the polarisation produced in calcium fluoride by an electric field of 500 V/m if it has a relative permittivity of 7.4?

Answer

Using equation [3]

$$P = \varepsilon_0 E(\varepsilon_r - 1) = 8.85 \times 10^{-12} \times 500 \times (7.4 - 1)$$
$$= 2.8 \times 10^{-8} \ C/m^2$$

Mechanisms of polarisation

There are four mechanisms by which polarisation can be produced in a dielectric.

1 *Electronic polarisation* When an electric field is applied to an atom the arrangement of the electrons in orbit about the nucleus is distorted with the electrons becoming more concentrated on one side of the nucleus. The result is a temporary dipole, the effect vanishing when the electric field is removed.

2 *Ionic polarisation* An ionic bonded material, such as sodium chloride, consists of an orderly array of positive and negative ions and since their arrangement is completely symmetrical there is no dipole. However, when an ionic bonded material is placed in an electric field forces act on the ions, pulling the positive ions in one direction and the negative ions in the opposite direction. The result is distortion and temporary dipoles are produced, the effect vanishing when the electric field is removed.

3 *Molecular polarisation* Some molecules have a non-symmetrical arrangement of electrons and so are permanent dipoles, e.g., water. Generally the directions of the dipoles are random in the absence of an electric field. The effect of an electric field is to line up these permanent dipoles.

4 *Interfacial polarisation* In real dielectrics the arrangement of the atoms, ions or molecules is not perfect and there are inevitably some gaps in the array and impurities present. In addition there can be a few free electrons. When an electric field is applied these may move through the material to its surfaces. The result is that the surface of the material nearest the positive plate can acquire a negative charge and the surface nearest the negative plate a positive charge.

In a dielectric such as polystyrene, a hydrocarbon, the predominant mode of polarisation is by electron displacement in atoms. Crystalline ceramics, such as alumina, are composed of bonded ions and the predominant mode of polarisation is ionic. Polar polymers such as polyester have polar groups of molecules and so the predominant mode of polarisation is molecular polarisation.

The effect of frequency

With an electric field applied across a dielectric the dipoles line up with the field. However if the electric field is alternating the dipoles have to keep switching directions. At low frequencies the dipoles have sufficient time to keep up with the changes in the electric field. However at high frequencies this is not the case; the dipoles cannot keep up with the changing field. The effect of this is a reduction in polarisation and hence a reduction in relative permittivity with increasing frequency.

The extent to which the dipoles cannot keep up with an alternating electric field depends on the mechanism by which they were produced. Interfacial polarisation is a very slow process and frequencies less than 1 Hz can be enough for this polarisation to vanish. At frequencies of the order of 10^4 to 10^6 Hz molecular polarisation vanishes as the molecular dipoles are unable to keep up with the changing field. At frequencies of the order of 10^9 to 10^{13} Hz ionic polarisation vanishes, and at about 10^{14} to 10^{16} Hz the electronic polarisation vanishes.

Dielectric loss

When a dielectric material is used in an alternating electric field a fraction of the energy is 'lost' each time the field is reversed. This loss is due mainly to what can be called the frictional effects involved in reversing the directions of the dipoles and current leakage. Dipole friction depends on the frequency and depends on the mechanism by which the polarisation has been produced. For a particular mechanism the greatest dielectric loss occurs at the frequency at which the polarisation due to that mechanism is about to be lost. Thus, for example, ionic polarised materials have losses which peak at a frequency in the region 10^9 to 10^{13} Hz and are low at lower frequencies. Microwave ovens have frequencies chosen so that for the material to be heated the dielectric losses are high but for the other materials the losses are low. Thus, for example, when microwave ovens are used to cure adhesive joints between sections of wood, the adhesive has a high loss factor and the wood a low loss factor at the microwave frequency.

Loss factor

The *loss factor* of a dielectric is defined as the ratio

$$\text{loss factor} = \frac{\text{energy lost per cycle}}{2\pi \times \text{maximum energy stored}} \qquad [4]$$

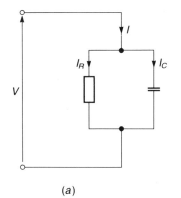
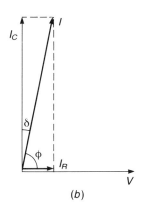

Fig. 3.4 (a) Equivalent parallel circuit for a lossy capacitor, (b) phasor diagram

(a)   (b)

This is generally expressed as the tangent of an angle, tan δ. The reason for this can be seen by considering the loss with an equivalent circuit for a real capacitor.

A capacitor with a dielectric loss can be represented by a loss-free capacitor with either a series or parallel resistor giving I^2R power dissipation. Figure 3.4 shows the resulting equivalent circuit and phasor diagram for a lossy capacitor represented by a resistor in parallel with a loss-free capacitor when an alternating voltage is applied. The current I_C through the capacitor arm of the circuit is 90° out of phase with the current I_R through the resistor arm. The current I is thus at an angle δ displaced from what it would have been if the capacitor had no losses. δ is called the *loss angle*. From the phasor diagram

$$\tan \delta = \frac{I_R}{I_C} = \left(\frac{V/R}{V/X_C} \right) = \frac{X_C}{R}$$

Since

$$X_C = \frac{1}{\omega C}$$

$$\tan \delta = \frac{1}{\omega RC} \tag{5}$$

The power loss in the parallel resistor is V^2/R, where V is the r.m.s. value. Thus the energy loss per cycle, i.e., the energy loss during a time of $2\pi/\omega$, is

$$\text{energy loss per cycle} = \frac{V^2}{R} \times \frac{2\pi}{\omega}$$

The maximum energy stored by the capacitor is $\frac{1}{2}CV_{max}^2$ (see Ch. 2), where V_{max} is $\sqrt{2}V$. Thus the loss factor is, by the definition given in equation [4],

$$\text{loss factor} = \frac{(2\pi V^2/R\omega)}{2\pi(\frac{1}{2}C2V^2)} = \frac{1}{\omega RC} = \tan \delta$$

The quantity $\tan \delta$ is the *loss factor* of the dielectric.

The power loss in the parallel resistor is V^2/R and so is, using the value of R given by equation [5],

$$\text{power loss} = V^2\omega C \tan \delta \qquad [6]$$

The circuit phase angle ϕ is related to the loss angle δ. Since ϕ is $(90° - \delta)$, then

$$\cos \phi = \cos (90 - \delta) = \sin \delta$$

For small loss angles $\sin \delta$ approximates to $\tan \delta$. Thus in this situation

$$\text{power factor} = \cos \phi \approx \tan \delta \qquad [7]$$

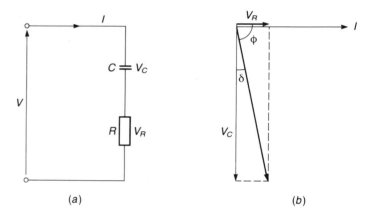

Fig. 3.5 (a) Equivalent series circuit for a lossy capacitor, (b) phasor diagram

(a)   (b)

The alternative to considering the lossy capacitor to be represented by a loss-free capacitor with a parallel resistor is as a loss-free capacitor in series with a resistor (Fig. 3.5). From the phasor diagram

$$\tan \delta = \frac{V_R}{V_C} = \frac{IR}{(I/\omega C)} = \omega RC$$

The power loss is I^2R and thus, using the value of R given in the above equation,

$$\text{power loss} = \frac{I^2 \tan \delta}{\omega C} \qquad [8]$$

In general, the representation of the losses by a parallel resistor is used for large capacitances while the series resistor is used for small capacitances.

Example 2

A capacitor, of capacitance 0.1 μF, has a loss factor of 0.003. What will be the power loss when it is connected across a 240 V, 50 Hz supply?

Answer

The loss factor is tan δ, hence representing the losses by a parallel resistor the power loss equation [6] gives

$$\text{power loss} = V^2 \omega C \tan \delta$$

$$= 240^2 \times 2\pi \times 50 \times 0.1 \times 10^{-6} \times 0.003$$

$$= 5.4 \times 10^{-3} \text{ W}$$

If the losses had been represented by the series resistor then equation [8] gives

$$\text{power loss} = \frac{I^2 \tan \delta}{\omega C}$$

and since $I = V\omega C$ then

$$\text{power loss} = \frac{(V\omega C)^2 \tan \delta}{\omega C} = V^2 \omega C \tan \delta$$

This is the same as equation [6] and so yields the same answer.

Example 3

What will be the size of the parallel resistor used in the equivalent circuit for a lossy capacitor if the capacitor has a capacitance of 0.5 μF, a loss factor of 0.002, and the frequency of the alternating field is 1 kHz?

Answer

Using the equation derived above

$$\tan \delta = \frac{1}{\omega RC}$$

Hence

$$R = \frac{1}{2\pi \times 1000 \times 0.5 \times 10^{-6} \times 0.002}$$

$$= 1.6 \times 10^5 \ \Omega$$

Dielectric properties

The following are some of the key properties of dielectric materials and determine their suitability for specific applications.

1 *Relative permittivity* This needs to be considered at the frequencies for which the material is to be used since the effect of frequency will depend on the mechanisms by

Table 3.1 Typical properties of commonly used dielectrics

Dielectric	Rel. permittivity		Dielectric strength 10^6 V/m	Volume resistivity Ω m	Loss factor tan δ
	50 Hz	10^6 Hz			
Air	1	1	3		
Alumina	9	6.5	6	10^9–10^{12}	0.0002–0.01
Glass (Pyrex)	4.3	4	14	10^{14}	0.01–0.02
Mica	7	7	40	10^{11}	0.0001
Paper, dry	2–3		16	10^{10}	0.001–0.008
Paraffin wax	2.3	2.3	10	10^{13}–10^{17}	0.003
Polyethylene	2.3	2.3	20	10^{13}–10^{16}	0.0002–0.0005
Polystyrene	2.5	2.5	20	10^{16}	0.0001–0.001
PTFE	2.1	2.1	20	10^{16}	0.0002
Titanium dioxide		100	6	10^{12}	0.0002–0.005
Barium titanate		3000	12	10^6–10^{13}	0.0001–0.02

Note The loss factor values quoted are typically those obtainable at about 1 MHz. All data are for ambient temperatures of about 20 °C.

which polarisation is produced in the dielectric. Relative permittivities of commonly used materials vary from about one to thousands, though for many the value tends to be in the region of 2 to 7. See Table 3.1.

2 *Dielectric strength* The dielectric strength is the maximum electric field, or potential gradient, that can be maintained in the dielectric (see Ch. 2). The dielectric strength depends on the material, its thickness, the shape and size of the conductors between which the dielectric is placed, moisture content of the material, temperature and pressure. See Table 3.1 for typical values.

3 *Surface and volume resistivity* Dielectrics need to have a high resistivity in order to prevent leakage of charge. The resistivity needs to be considered both for the volume of the material and its surfaces. Surfaces can acquire absorbed films of moisture and other contaminants and the surfaces of the dielectric can thus have a significantly lower resistivity than the volume of the dielectric.

4 *Temperature effects* One of the main effects of an increase in temperature is a decrease in resistivity and hence higher leakage currents through the dielectric. If the temperature increase is large enough this leakage current may become large enough to produce itself significant heating which further increases the leakage current which further increases the heating effect . . . and so on until this thermal runaway effect burns up the material when the heat is generated faster than it can be dissipated.

Other thermal properties that need to be considered in selecting a dielectric are melting points (e.g., for the wax used in impregnating paper), freezing points for liquid dielectrics, the thermal expansivity, specific heat capacity, flash-point or ignitability, and ageing effects.

5 *Dielectric losses* The extent to which a dielectric in an alternating electric field 'loses' energy is indicated by the loss factor, tan δ (see earlier this chapter). See Table 3.1 for typical values. A very good dielectric would have a loss factor of the order of 10^{-5} with a poor one about 0.1.

6 *Mechanical properties* In selecting a dielectric there may be a need to consider the mechanical properties of the material. Many are, for example, very brittle. Properties which may need to be considered are the effects of bending, impact strength, tearing strength, tensile strength, compressive strength, shear strength, and machinability.

Types of capacitors

The following are the forms and materials of commonly available capacitors.

1 *Mica capacitors* These generally consist of thin sheets of mica, about 2.5 μm thick, coated on both sides with silver (Fig. 3.6(a)). An alternative form is of mica sheets sandwiched between sheets of lead or aluminium foil. Mica capacitors are low loss, good for radio frequencies, have good temperature stability and have values up to about 1000 pF. They do however tend to be rather bulky.

2 *Paper capacitors* These consist of layers of waxed paper sandwiched between layers of metal foil, the whole being

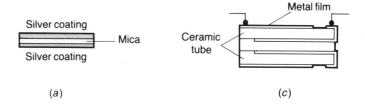

(a)   (c)

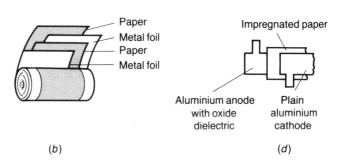

Fig. 3.6 (a) Mica capacitor, (b) paper capacitor, (c) ceramic tubular capacitor, (d) electrolytic capacitor

(b)   (d)

wound into a roll (Fig. 3.6(*b*)). Such capacitors tend to be used where losses are not too important, and have a capacitance which changes significantly with temperature, a working life which is shorter than other capacitors and values between about 500 pF and 10 μF. Working voltages can be up to 150 kV.

3 *Plastic capacitors* These have a similar construction to paper capacitors, consisting of layers of a plastic film, e.g., polystyrene or PTFE, between layers of metal foil. They are very reliable capacitors, maintaining their capacitance values at high temperatures and over long periods of time.

4 *Ceramic capacitors* These can be in tube (Fig. 3.6(*c*)) or plate forms, the form of construction depending on the capacitance value required. Ceramic materials have high relative permittivities and so the capacitors can be very compact. They have high working voltages, can be used at high frequencies, and have capacitances between about 1 pF and 0.1 μF.

5 *Electrolytic capacitors* These consist of aluminium foil plates separated by a thick absorbent material, e.g., paper, impregnated with an electrolyte such as ammonium borate (Fig. 3.6(*d*)). Electrolytic action occurs when a potential difference is connected between the plates and results in a thin layer of aluminium oxide being formed on the positive plate. This very thin layer forms the dielectric. Another form of electrolytic capacitor uses tantalum instead of aluminium, with tantalum oxide forming the dielectric. The electrolytic type of capacitor must always be used with a d.c. supply and must always be connected with the correct polarity. For their size such capacitors, because of the thinness of the dielectric, have very high capacitances with values tending to be in microfarads. Working voltages tend to be between about 6 V and 600 V.

Example 4

What is the total area of dielectric needed for a paper capacitor of capacitance 0.01 μF and its breakdown voltage if the paper has a thickness of 1 mm, a dielectric strength of 1.6×10^7 V/m and a relative permittivity of 2.5?

Answer

Using the equation for the capacitance of a parallel plate capacitor (equation [5] Ch. 2)

$$C = \frac{\varepsilon_r \varepsilon_0 A}{d}$$

$$A = \frac{0.01 \times 10^{-6} \times 0.001}{2.5 \times 8.85 \times 10^{-12}} = 0.45 \text{ m}^2$$

The dielectric strength is the breakdown voltage divided by the dielectric thickness. Hence

$$V = 1.6 \times 10^7 \times 0.001 = 1.6 \times 10^4 \text{ V}$$

Electrical insulators

The properties required of a good electrical insulator are generally that it should have a high surface and volume resistivity, a high dielectric strength, a low loss factor, but not a high dielectric constant. The high resistivity is to reduce leakage currents. The high dielectric strength is to prevent breakdown of the insulation at high voltages. The dielectric constant should be small so as to reduce polarisation and so prevent the build-up of charge on the insulator.

Piezo-electricity

When an electric field is applied to a dielectric polarisation occurs. This distortion of atoms and molecules, and rotation of permanent dipoles cause dimensional changes in the solid dielectric. Thus an electric field produces a strain in the material. Strain is defined as the change in length per unit length of material. This change in dimensions is called *electrostriction*. It occurs with all dielectrics but is usually very small unless a very high electric field is used.

With some dielectric materials, when they are stretched or compressed and a dimensional change is produced then one face of the material becomes positively charged and the opposite face negatively charged, i.e. an electric field has been produced in the material. This effect is called *piezo-electricity*. Piezo-electric materials consist of ions in an orderly array within a crystal. With many crystals when the crystal is stretched or squashed and the separation of the ions changed no piezo-electric effect is produced. This is because the arrangement of positive and negative ions in such crystals is completely symmetrical and the change in the separation of ions produced by the applied forces does not result in any dipole moment. However, when the arrangement of the positive and negative ions is not symmetrical, a dipole moment is produced. Typical piezo-electric materials are quartz and barium titanate.

For a piezo-electric material, the electric field E produced by a stress σ is proportional to the stress. Stress is defined as the force applied per unit area of material.

$$E = g\sigma \hspace{4cm} [9]$$

where g is a constant, sometimes called the *voltage output coefficient*. Conversely, for a piezo-electric material subject to an electric field E a strain ε is produced with ε being proportional to E.

Table 3.2 Typical values of the piezo-electric constant

Material	Piezo-electric constant d in 10^{-12} m/V
Quartz	2.3
Barium titanate	100
Lead zirconate titanate (PZT)	250
Ammonium dihydrogen phosphate (ADP)	50

$$\varepsilon = dE \qquad\qquad [10]$$

where d is a constant, sometimes called the *piezo-electric constant (or modulus or coefficient)*. Typical values of the piezo-electric constant are shown in Table 3.2.

The strain produced by the application of the electric field is exactly the same as an applied stress would need to produce to give the electric field in the converse effect. Thus the constants d and g are related.

The strain ε produced by a stress σ is given by

$$\text{modulus of elasticity} = \frac{\sigma}{\varepsilon}$$

Thus electric field $E = g\sigma = g(\varepsilon \times \text{modulus of elasticity})$. Since $\varepsilon = dE$, then

$$\text{electric field } E = \frac{\varepsilon}{d} = g(\varepsilon \times \text{modulus of elasticity})$$

$$\text{modulus of elasticity} = \frac{1}{gd} \qquad\qquad [11]$$

Piezo-electric materials are widely used as transducers. For example, they are used in the pickup of record players to convert the movement of the needle imposed by the grooves on the record into electrical signals. This movement flexes a strip of crystal and thus stresses it, hence the production of a potential difference across its faces which is related to the movement of the needle. Another application is for the generation and detection of ultrasonic waves, i.e., high-frequency pressure waves. When an alternating potential difference is applied across the faces of a crystal an alternating strain is produced, i.e., the separation of the faces of the crystal increases and decreases. The result of this movement of the crystal faces is an alternating pressure wave in the surrounding medium and thus an ultrasonic wave generator. If such a wave is incident on a piezo-electric crystal then it becomes subject to an alternating strain and hence produces an alternating potential difference between its faces, hence an ultrasonic wave detector.

Example 5

What is the potential difference produced across a barium titanate wafer, 0.20 mm thick, if it is subject to a compressive stress of 30 MPa? The piezo-electric constant for barium titanate is 1.0×10^{-10} m/V and the modulus of elasticity 70 GPa.

Answer

The electric field E produced by a stress σ is given by equation [9] as

$$E = g\sigma$$

The modulus of elasticity is related to the constants g and d by equation [11] as

$$\text{modulus of elasticity} = \frac{1}{gd}$$

Hence

$$E = \frac{\sigma}{d(\text{modulus of elasticity})}$$

The electric field E is the potential gradient across the wafer, i.e., V/thickness. Hence

$$V = \frac{30 \times 10^6 \times 0.0002}{1.0 \times 10^{-10} \times 70 \times 10^9}$$

$$= 857 \text{ V}$$

Crystal-controlled oscillators

An important use of quartz crystals is for stabilising the frequency of oscillators. When an alternating potential difference is applied between opposite faces of the crystal (Fig. 3.7) the crystal faces are set into vibration. Pressure waves travel through the crystal. Thus we can think of the alternating displacement of one crystal face resulting in a wave which travels through the crystal to the opposite face. If when it arrives at the opposite face it meets up with that face moving in the same direction as a result of the alternating potential difference then resonance occurs. For this to happen the distance L travelled in the crystal must be half a wavelength, or a multiple of half a wavelength, i.e.,

$$L = \tfrac{1}{2}n\lambda$$

where n is an integer. Since $v = f\lambda$, where v is the velocity in the crystal (note that the velocity in a crystal depends on the

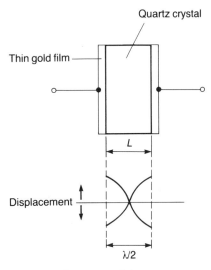

Fig. 3.7 Quartz crystal resonance condition

direction through the crystal in which the wave is travelling) and f the frequency of the vibration, then

$$f = \frac{nv}{2L} \qquad [12]$$

When $n = 1$ the frequency is the fundamental frequency of vibration; higher n values give harmonics. The maximum amplitude of the mechanical vibrations of the crystal faces occurs when the applied alternating potential difference frequency is in resonance with the mechanical vibrations.

When a quartz crystal is included in an oscillator circuit it will respond to an alternating potential difference across it; however because of the very high Q-factor of quartz crystals (typically between 10 000 and 500 000) the response is very significant only for frequencies very close to the resonant frequency. A vibrating crystal produces an alternating potential difference across its faces and so the impedance of the crystal is a minimum at this frequency and can be used to govern the frequency at which the oscillator circuit oscillates. The resonant frequency of a quartz crystal is affected by temperature changes and thus for accurate control the crystal is maintained at a constant temperature in a small oven.

Example 6

If the velocity of pressure waves in a quartz crystal in a particular direction is 5.5×10^3 m/s, what will be the fundamental frequency of vibration of a 2.0 mm thick crystal?

Answer

The fundamental frequency occurs when the thickness is ½λ and so the thickness is

$$\text{thickness} = \frac{v}{2f}$$

Hence the fundamental frequency is

$$f = \frac{5.5 \times 10^3}{2 \times 2.0 \times 10^{-3}}$$

$$= 1.4 \text{ MHz}$$

Ferroelectricity

With dielectrics, when an electric field is applied polarisation occurs and dipoles become aligned with the field. When the field is removed, even when the material has permanent dipoles, the polarisation generally vanishes. This is because the dipoles become randomly orientated and the effects of the dipoles cancel each other out. There are however some

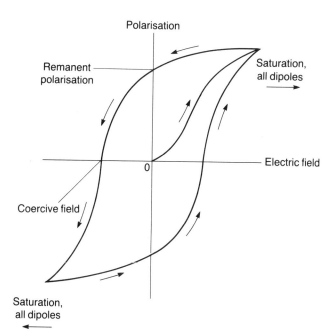

Fig. 3.8 The ferroelectric hysteresis loop

materials, called *ferroelectrics* for which this is not the case. When the electric field is removed the material still retains some polarisation.

Figure 3.8 shows what happens with a ferroelectric material. Initially when the electric field strength is increased, from point 0, the polarisation increases as more and more of the dipoles line up. Eventually the field is high enough for saturation to occur when all dipoles are lined up with the field. When the field is decreased from this point the polarisation decreases but does not reach zero when the field is zero, there being *remanent polarisation*. When the direction of the electric field is reversed this remanent polarisation decreases until at the *coercive field* value the polarisation becomes zero. Further increase of the field in this reverse direction enables a saturation point to be reached. When this field is decreased from this point the sequence just reverses itself. The result is a ferroelectric *hysteresis loop*.

Ferroelectric materials are always piezo-electric materials, but not all piezo-electric materials are ferroelectrics. Ferro-electricity is only maintained for a specific material to a particular temperature, called the *Curie point*. At temperatures above this the material is no longer ferroelectric. Figure 3.9 shows how the relative permittivity of a ferroelectric varies with temperature, there being a sharp increase to a very high value just before the Curie point and then at the point a very steep drop. For example, barium titanate has a relative permittivity of the order of 2000 at about room temperature,

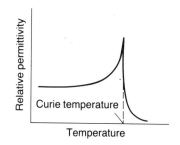

Fig. 3.9 The effect of temperature on ferroelectrics

Table 3.3 Examples of ferroelectrics and their Curie temperatures

Material	Curie temperature °C
Strontium titanate	−200
Rochelle salt	24
Barium titanate	120
Potassium niobate	435
Lead metaniobate	570

with a sharp increase to about 7000 at the Curie temperature of 120° C. Table 3.3 shows some of the common ferroelectric materials and their Curie point temperatures.

Liquid crystals

In general liquids have no particular order among their molecules. Whereas in a solid crystal the atoms can be thought of as being packed in an orderly way, in the liquid the molecules can be thought of as just tossed into a random pile with constant chaotic rearrangement. However, some liquids have rod-shaped molecules which in the liquid state are able to take up defined orientations with respect to each other and to any solid interface with the liquid and thus assume an orderly packing (Fig. 3.10). In the liquid the rod molecules align themselves all in the same direction. At a solid–liquid interface the rod molecules can be either at right-angles to the interface or parallel to it. Because of this order they are referred to as *liquid crystals*.

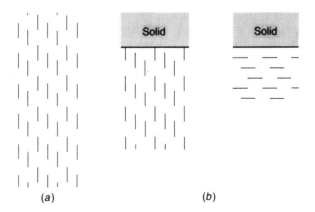

Fig. 3.10 Liquid crystals, (a) arrangement of molecules within the liquid, (b) at solid–liquid interfaces

If an electric field is applied to a liquid crystal then, provided it is above some critical value, the molecules within the liquid will align themselves with the field (Fig. 3.11). It is this ability to cause the molecules to change alignment directions that is the basis of the liquid-crystal displays used in digital watches, calculators and some instrument displays.

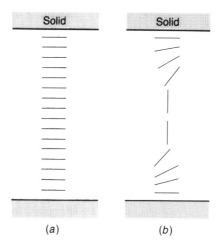

Fig. 3.11 Effect of an electric field on a liquid crystal, (a) no field, (b) field above the critical value

(a)   (b)

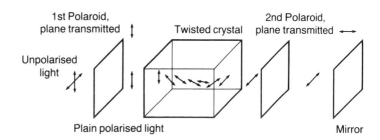

Fig. 3.12 Liquid crystal display

Figure 3.12 shows the form of such a liquid-crystal display. The light incident on the display is unpolarised, i.e., it has its planes of vibrations in all directions. This is made plane polarised, i.e., a plane of vibration in just one plane, by passing it through a sheet of Polaroid. This light is then incident on a liquid crystal which has its walls twisted. The effect of this is to cause the rod molecules within the crystal to follow an ordered twisted line through the crystal so that they rotate through 90°. The plane polarised light in passing through this twisted array also becomes twisted and emerges with its plane of polarisation rotated through the 90°. The light then passes through a second Polaroid which is so aligned as to transmit this rotated plane of polarisation. Following reflection at a mirror the light retraces its path. Thus in this condition light emerges from the crystal and it appears bright. However, if an electric field above the critical value is applied the molecules can be caused to become realigned so that the light does not have its plane of polarisation rotated on passing through the crystal and so cannot be transmitted through the second Polaroid. As a consequence no light then emerges from the crystal and it appears dark. By using a number of such liquid crystals in a dot-matrix form or as segments which can be used to constitute numbers or letters the application of suitable potential differences across them can be used to generate bright numbers and letters or other displays.

Problems

1 Explain what is meant by polarisation when applied to a dielectric and the various mechanisms by which polarisation can be produced.

2 What is the polarisation in sodium chloride in an electric field of 400 V/m if sodium chloride has a relative permittivity of 5.9?

3 What is the physical basis for energy loss in a dielectric when it is subject to an alternating electric field?

4 Derive an expression for the power loss in a capacitor dielectric when an alternating potential difference is applied by using an equivalent circuit model for a lossy capacitor.

5 Calculate the power loss in a 2 μF capacitor when it is connected to a 1 kV, 50 Hz supply and it has a loss angle of 0.020 rad.

6 Calculate the resistance of the parallel resistor for the parallel equivalent circuit for the lossy capacitor specified in problem 5.

7 A mica capacitor is to be made of mica of thickness 2.5 μm. What will be the maximum allowable voltage for the capacitor? The dielectric strength of mica can be taken as 4.0×10^7 V/m.

8 What is the capacitance of a parallel plate capacitor which has a sheet of PTFE 10 mm $\times$ 10 mm $\times$ 0.04 mm sandwiched between layers of conducting foil if the PTFE has a relative permittivity of 2.1?

9 What dielectric properties are required for a high-voltage, low-loss, compact capacitor?

10 Explain what the properties of piezo-electricity and ferroelectricity mean for a material.

11 What potential difference will be produced across a 0.02 mm thick slice of quartz when it is subject to a stress of 15 MPa if quartz has a tensile modulus of 70 GPa and a piezo-electric constant of 2.3×10^{-12} m/V?

12 If a ferroelectric material has a remanent polarisation of 7.0×10^{-8} C/m^2 and a coercive field of 3000 V/m, what will be (a) the voltage that has to be applied across a 0.10 mm thick layer to reduce the polarisation to zero and (b) the residual polarisation when the field is zero?

4 Magnetic fields

Introduction

The approach to magnetic fields that has been adopted in this chapter is, following some preliminary discussion of fields, to start with electromagnetic induction and use it to define the term flux. From this definition the concept of flux leads to a consideration of flux density, magnetomotive force, magnetic field strength, magnetic circuits and inductance. An alternative approach would have been to start with a definition of flux density in terms of the force on a current-carrying conductor in a magnetic field and then derive electromagnetic induction from it. In this chapter the force on a current-carrying conductor derives from electromagnetic induction and the concept of flux. This approach has been adopted because of the importance of the flux concept in the study of magnetic circuits and inductance, the primary concerns of this chapter.

Magnetic fields

In the vicinity of permanent magnets and current-carrying elements a magnetic field is said to exist. The existence of such a field can be established by forces acting on other permanent magnets or current-carrying elements when they are placed in the field. The direction of the field at a point is the direction of the force acting on an imaginary north pole placed at that point.

The magnetic-field pattern in the space surrounding a permanent magnet or a current-carrying conductor can be plotted using a compass needle, the needle at each point lining up with the direction of the field at that point, or demonstrated by scattering iron filings in the vicinity. The types of patterns that are produced are shown in Fig. 4.1. In the case of current-carrying wires a useful way of remembering the direction of the field round a wire is the *corkscrew rule*. If a right-handed corkscrew is driven along a wire in the direction of the current in that wire then the corkscrew rotates in the direction of the magnetic field.

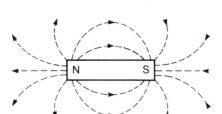

(a)

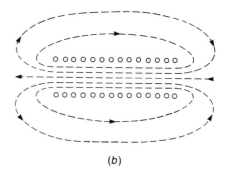

(b)

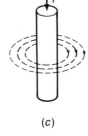

(c)

Fig. 4.1 Magnetic field patterns for (a) a bar magnet when the earth's magnetic field is insignificant, (b) a current-carrying solenoid, (c) a current-carrying wire

Magnetic flux

The term *magnetic line of force* is used for a line traced out by the plotting of field patterns. For a permanent magnet it represents the direction a free north pole would move in moving from one point on the magnet to another. All the lines appear to emerge from a region in the magnet called the north pole and converge on a region called the south pole. With the current-carrying solenoid the lines of force form closed loops which, if we imagined one end of the solenoid to be a north pole and the other a south pole, emerge from a north pole and converge on a south pole. Within the solenoid the lines continue between the poles.

A useful way of considering magnetic fields is in terms of magnetic flux (Φ). This is considered to flow out of north poles and into south poles. The lines of force are lines along which magnetic flux flows. In the case of the permanent magnet the flux lines pass outside the magnet from the north pole to the south pole and are considered to complete a loop by passing through the material of the magnet (Fig. 4.2), producing a pattern just like that of the current-carrying solenoid.

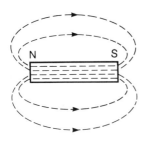

Fig. 4.2 Lines of magnetic flux with a permanent magnet

Electromagnetic induction

Magnetic fields can be described by means of the lines of magnetic flux. A quantitative measure of magnetic flux can be obtained by means of electromagnetic induction. *Electromagnetic induction* is said to occur when the magnetic flux linked by a circuit changes, the consequence of such change

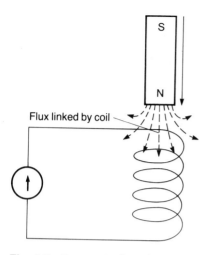

Flux linked by coil

Fig. 4.3 Demonstration of electromagnetic induction

being the production of an e.m.f. in the circuit. The term 'linked' can be explained by considering the magnetic flux to be streaming out along the lines of force like water from a hosepipe, some flux being linked by the circuit if some of the 'water from the hosepipe' passes through the loop formed by the wires of the circuit.

Electromagnetic induction can be demonstrated by the simple apparatus shown in Fig. 4.3 in which a magnet is moved towards a coil of wire. An e.m.f. is induced in the coil of wire when the magnetic flux linked by it changes. The size of the induced e.m.f. is proportional to the rate of change of the magnetic flux linked by the coil. The direction of the resulting currents in the coil is always in such a direction as to set up magnetic fields which tend to neutralise the change in magnetic flux linked by the coil which caused them.

We can define the relationship between the e.m.f. e induced in a turn of wire in the coil and the rate of change of flux linked by that turn, i.e., $d\Phi/dt$, as

$$e = -\frac{d\Phi}{dt} \qquad [1]$$

The minus sign indicates that the induced e.m.f. is in such a direction as to oppose the change causing it. The unit of flux is the weber (Wb). If the flux linked changes by one weber per second then the induced e.m.f. is 1 volt.

For a coil of N turns, each turn will produce an induced e.m.f., so the total e.m.f. induced will be the sum of those due to each turn. Hence

$$e = -N\frac{d\Phi}{dt} \qquad [2]$$

Flux density

The term *magnetic flux density B* is used for the amount of flux passing through unit area. Thus if flux Φ passes through an area A then

$$B = \frac{\Phi}{A} \qquad [3]$$

The unit of flux density is the tesla (T) when the flux is in webers (Wb) and the area in square metres.

Magnetomotive force

Increasing the current through a conductor increases the magnetic field around it. We can thus consider the magnetic flux produced as a result of the current to be proportional to the current. With an electrical circuit the source of current, i.e., charge flow, is the electromotive force (e.m.f.). With magnetism we talk of the source of magnetic flux being the *magnetomotive force (m.m.f.)*. The magnetomotive force of a coil through which a current flows is proportional to the current. We can think of each turn of the coil being responsible for producing flux and thus the total magnetomotive force for a coil depends on the number of turns N. The magnetomotive force is thus defined as

$$\text{m.m.f.} = NI \qquad [4]$$

The unit of m.m.f. is the ampere. However because the m.m.f. is the product of number of turns and the current it is sometimes stated as ampere-turns.

Example 1

What is the m.m.f. produced by a current of 2 A passing through a coil of 500 turns?

Answer

Using equation [4]

$$\text{m.m.f.} = NI = 500 \times 2 = 1000 \text{ A}$$

Magnetic field strength

With an electrical circuit an e.m.f. drives a current through conductors. With magnetism the m.m.f. produces flux which flows through the lines of flux. The equivalent in magnetism of the potential difference per unit length of conductor in the electrical circuit is the m.m.f. per unit length of flux path. This is given a name, the *magnetic field strength (H)*. It is also known as the *magnetic field intensity* or the *magnetising force*.

$$H = \frac{\text{m.m.f.}}{L} = \frac{NI}{L} \qquad [5]$$

The unit of magnetic field strength is amp/metre.

Example 2

What is the magnetic field strength at the centre of a solenoid of 1000 turns and length 200 mm if it carries a current of 5 A?

Answer

Using equation [5]

$$H = \frac{NI}{L} = \frac{100 \times 5}{0.200} = 25\,000 \text{ A/m}$$

Permeability

Inside a current-carrying solenoid there is a magnetic field which we can consider to be the result of magnetic flux passing through the solenoid in lines parallel to its axis (see Fig. 4.1(*b*)). This flux is related to the m.m.f. of the coil. Hence the flux density is related to the magnetic field strength H, since $H = \text{m.m.f.}/L$, where L is the length of the coil. The flux density in the solenoid also depends on what medium is inside the solenoid, e.g., a solenoid consisting of wire wrapped round a piece of iron gives, for the same number of turns and current, a much stronger electromagnet than a solenoid with only air inside the turns. Thus B is related to H by

$$B = \mu H \qquad [6]$$

where μ is called the *absolute permeability* of the medium. If there is a vacuum the permeability is called the *permeability of free space* (μ_0) and has the value $4\pi \times 10^{-7}$ T/A or H/m (henrys per metre). For other mediums the absolute permeability can be expressed as a multiple of the μ_0.

Absolute permeability = relative permeability $\mu_r \times \mu_0$

$$\mu = \mu_r \mu_0 \qquad [7]$$

The flux density B_0 produced in a vacuum for a given magnetic field strength H is B_0/μ_0. The flux density B produced in some material for the same magnetic field strength H is $B/(\mu_0 \mu_r)$. Thus

$$\frac{B_0}{\mu_0} = \frac{B}{\mu_0 \mu_r}$$

Hence the relative permeability is

$$\mu_r = \frac{B}{B_0} = \frac{\text{flux density in the material}}{\text{flux density in a vacuum}} \qquad [8]$$

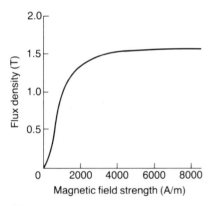

Fig. 4.4 *B–H* relationship for cast steel

The relative permeability thus specifies by what factor the flux density in a material is increased when compared with a vacuum. Since air behaves much the same as a vacuum, the relative permeability of air is 1.

The relative permeability for a particular material is not generally the same for all values of magnetic field strength (or flux density). Figure 4.4 shows the type of relationship that exists between the flux density and magnetic field strength for a typical magnetic material. The effect on the value of relative permeability is shown in the derived relative permeability against magnetic field strength graph, Fig. 4.5. Thus in solving problems involving relative permeability care has to be exercised in assuming a single, constant, value for the relative permeability. Often recourse has to be made to tables of data or graphs similar to Figs 4.4 and 4.5.

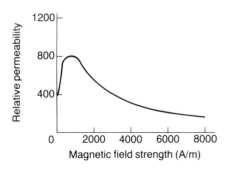

Fig. 4.5 Variation of relative permeability with field strength for cast steel

Example 3

A solenoid with 600 turns has a length of 200 mm and a cross-sectional area of 2.0 cm². If the current through the solenoid is 3.0 A what will be (*a*) the magnetic field strength, (*b*) the flux density if it is an air solenoid, (*c*) the flux density if the solenoid core is a material with a relative permeability of 500?

Answer

(*a*) Using equation [5]

$$H = \frac{NI}{L} = \frac{600 \times 3.0}{0.200} = 9000 \text{ A/m}$$

(*b*) With a vacuum, the flux density is given by equation [6] as

$$B = \mu_0 H = 4\pi \times 10^{-7} \times 9000 = 0.011 \text{ T}$$

(*c*) With the material of relative permeability 500, the flux density will be just 500 times greater than the value in the vacuum.

$$B = \mu_r \mu_0 H = 500 \times 4\pi \times 10^{-7} \times 9000 = 5.5 \text{ T}$$

Magnetic circuit

For the simple electrical circuit shown in Fig. 4.6(*a*) there is a source of e.m.f. which drives a current through the circuit resistance. Figure 4.6(*b*) shows the magnetic equivalent of this simple circuit. It consists of a source of m.m.f. which drives flux through 'magnetic resistance'. The term used for magnetic resistance is *reluctance* (*S*).

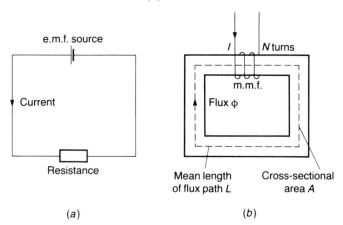

Fig. 4.6 (*a*) Electrical circuit and (*b*) equivalent magnetic circuit

(*a*) (*b*)

For the electrical circuit

e.m.f. = *IR*

where *R* is the total resistance of the circuit.

For the magnetic circuit

m.m.f. = Φ*S* [9]

where S is the total reluctance of the circuit. Reluctance has the unit A/Wb.

For the magnetic circuit shown, the source of m.m.f. is a coil with N turns and carrying a current I. Thus the m.m.f. is NI. This produces a flux Φ in the core and hence the flux density B produced in the core is Φ/A, where A is the cross-sectional area of the core. But (equation [6])

$$B = \mu_r\mu_0 H$$

where H is the magnetic field strength in the core and μ_r is the relative permeability of the core. But $H = $ m.m.f.$/L$, where L is the length of flux path. Hence

$$\Phi = BA = \mu_r\mu_0 HA = \mu_r\mu_0(\text{m.m.f.}/L)A$$

Hence the reluctance of the core S is given by equation [9] as

$$S = \frac{\text{m.m.f.}}{\Phi} = \frac{\text{m.m.f.}}{\mu_r\mu_0(\text{m.m.f.}/L)A}$$

$$S = \frac{L}{\mu_r\mu_0 A} \qquad [10]$$

This equation can be compared with the equivalent relationship for the resistance R of an electrical conductor,

$$R = \frac{\varrho L}{A}$$

where ϱ is the resistivity, A the cross-sectional area and L the length of the conductor.

With electrical circuits the reciprocal of resistance is called the conductance. With magnetic circuits the reciprocal of reluctance is called the *permeance*.

Example 4

What is (*a*) the reluctance and (*b*) the flux in the core of a magnetic circuit if the flux path has a mean length of 200 mm and the core is made of material with a uniform cross-sectional area of 100 mm^2 and relative permeability 50, the m.m.f. being provided by a coil of 50 turns of wire carrying a current of 200 mA wound on the core?

Answer

(*a*) The magnetic circuit will be of the form shown in Fig. 4.6(*b*). The reluctance S is given by equation [10] as

$$S = \frac{L}{\mu_r\mu_0 A} = \frac{0.200}{50 \times 4\pi \times 10^{-7} \times 100 \times 10^{-6}} = 3.2 \times 10^7 \text{ A/Wb}$$

(*b*) For the magnetic circuit, equations [4] and [9] give

$$\text{m.m.f.} = NI = \Phi S$$

$$\Phi = \frac{NI}{S} = \frac{50 \times 0.200}{3.2 \times 10^7} = 3.1 \times 10^{-7} \text{ Wb}$$

Laws for magnetic circuits

The equivalent for magnetic circuits of Kirchhoff's laws for electrical circuits is:

1　The total magnetic flux entering a junction equals the flux leaving the junction.
2　In any closed magnetic circuit, the algebraic sum of the product of the magnetic field strength and the length of flux path for each part of the circuit is equal to the resultant magnetomotive force (m.m.f.).

Series- and parallel-connected magnetic circuits

Figure 4.7 shows a simple electrical circuit with two resistors in series. Since the potential difference V across the two resistors is the sum of the potential differences across each, then

$$V = V_1 + V_2$$

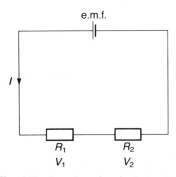

Hence, since the resistors are in series the current through each resistor is the same and so

$$IR = IR_1 + IR_2$$

The total resistance R of two resistors R_1 and R_2 in series is thus

$$R = R_1 + R_2$$

Fig. 4.7 A series electrical circuit

Hence in the absence of any other resistance in the circuit

$$\text{e.m.f.} = IR = I(R_1 + R_2)$$

Figure 4.8 shows a simple magnetic circuit with two reluctances in series. The m.m.f. is the magnetic equivalent of the electrical potential difference, thus the m.m.f across the two reluctances is given by

$$\text{m.m.f.} = \text{m.m.f.}_1 + \text{m.m.f.}_2$$

Because the reluctances are in series the same flux will flow through each. Thus

$$\Phi S = \Phi S_1 + \Phi S_2$$

Thus, in the same way as with the electrical circuit, the total reluctance S of two reluctances S_1 and S_2 is

$$S = S_1 + S_2 \qquad [11]$$

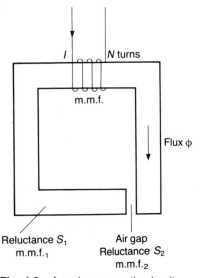

Fig. 4.8 A series magnetic circuit

Hence in the absence of any other resistance in the circuit

$$\text{e.m.f.} = IR = I(R_1 + R_2)$$

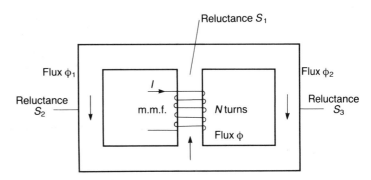

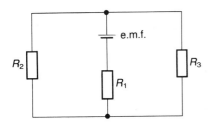

Fig. 4.9 A parallel magnetic circuit

Fig. 4.10 The equivalent parallel
electrical circuit

Figure 4.9 shows a parallel magnetic circuit, with Fig. 4.10 the equivalent parallel electrical circuit. The electrical circuit consists of a resistance R_1 in series with the source of e.m.f. and two resistors R_2 and R_3 in parallel. For the two resistors in parallel, their combined resistance R_p is given by

$$\frac{1}{R_p} = \frac{1}{R_2} + \frac{1}{R_3}$$

$$R_p = \frac{R_2 R_3}{R_2 + R_3}$$

The circuit can now be considered to consist of R_p in series with R_1 and so the total circuit resistance R is

$$R = R_p + R_1$$

Hence since e.m.f. $= IR$, then

$$\text{e.m.f.} = I\left(\frac{R_2 R_3}{R_2 + R_3} + R_1\right)$$

The magnetic circuit can be treated in exactly the same way. There is a reluctance of S_1 in series with the m.m.f. and two parallel reluctances of S_2 and S_3. For two reluctances in parallel, the total flux Φ through the parallel arrangement is the sum of the fluxes through each element.

$$\Phi = \Phi_1 + \Phi_2$$

But the m.m.f. across each reluctance will be the same since they are in parallel. Hence

$$\frac{\Phi}{\text{m.m.f.}} = \frac{\Phi_1}{\text{m.m.f.}} + \frac{\Phi_2}{\text{m.m.f.}}$$

Thus the combined reluctance of the two parallel elements S_p is given by

$$\frac{1}{S_p} = \frac{1}{S_2} + \frac{1}{S_3} \qquad [12]$$

The total reluctance of the circuit S is thus S_p in series with S_1, i.e. $S_p + S_1$. Hence since m.m.f. $= \Phi S$, then

$$\text{m.m.f.} = \Phi \left(\frac{S_2 S_3}{S_2 + S_3} + S_1 \right)$$

Example 5

For the magnetic circuit shown in Fig. 4.11, what is the flux in the air gap when a current of 1.2 A passes through the coil of 800 turns? The core has a relative permeability of 50 and air 1.

Answer

The reluctance of the core is given by equation [10] as

$$S = \frac{L}{\mu_r \mu_0 A} = \frac{34.5 \times 10^{-2}}{50 \times 4\pi \times 10^{-7} \times 4 \times 10^{-4}}$$

$$= 1.3 \times 10^7 \text{ A/Wb}$$

The reluctance of the air gap is given by

$$S = \frac{L}{\mu_r \mu_0 A} = \frac{0.005}{1 \times 4\pi \times 10^{-7} \times 4 \times 10^{-4}}$$

$$= 9.9 \times 10^6 \text{ A/Wb}$$

The total reluctance is the sum of the core and air gap reluctances since the two are in series (equation [11]). Hence

$$\text{total reluctance} = 2.3 \times 10^7 \text{ A/Wb}$$

The flux through the air gap will be the same as the flux through the core and hence the entire circuit. This is because they are in series. Hence the flux Φ is given by m.m.f. $= \Phi S$ and so

$$\Phi = \frac{\text{m.m.f.}}{S} = \frac{NI}{S} = \frac{800 \times 1.5}{2.3 \times 10^7} = 5.2 \times 10^{-5} \text{ Wb}$$

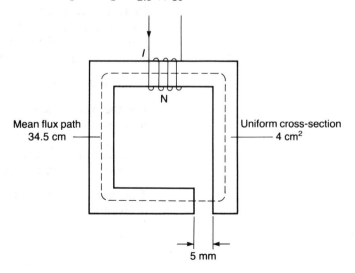

Fig. 4.11 Example 5

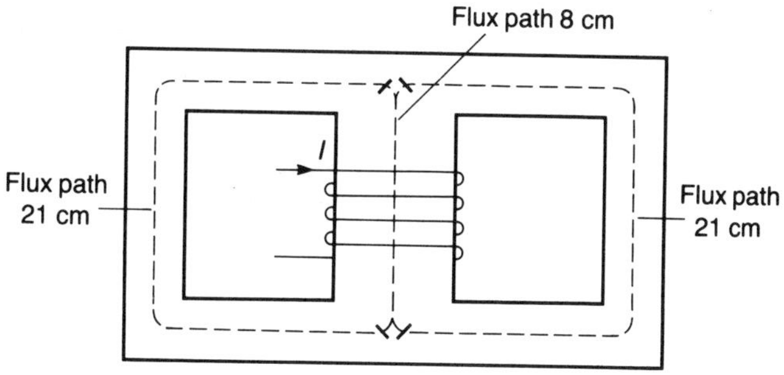

Fig. 4.12 Example 6

Example 6

For the magnetic circuit shown in Fig. 4.12 what will be the current needed in the 600-turn coil to produce a flux of 1.5×10^{-4} Wb in the central arm? The material has a relative permeability of 200 and the thickness of the entire core is 2 cm.

Answer

The two outer arms are the same size and material. Hence their reluctance is given by equation [10] as

$$S = \frac{L}{\mu_r \mu_0 A} = \frac{21 \times 10^{-2}}{200 \times 4\pi \times 10^{-7} \times 4 \times 10^{-4}}$$

$$= 2.09 \times 10^6 \text{ A/Wb}$$

The length of the flux path has been taken as the length of the central axis of the arms.

Because the two outer arms have the same reluctance and are in parallel their combined reluctance is half the reluctance of one arm (equation [12]), i.e.,

$$S = 1.0 \times 10^6 \text{ A/Wb}$$

The central core has a reluctance of

$$S = \frac{L}{\mu_r \mu_0 A} = \frac{8 \times 10^{-2}}{200 \times 4\pi \times 10^{-7} \times 8 \times 10^{-4}}$$

$$= 4.0 \times 10^5 \text{ A/Wb}$$

The total reluctance of the circuit is the sum of the central arm and the combined outer arm reluctances, thus

total reluctance $= 1.4 \times 10^6$ A/Wb

The flux in the central arm of 1.5×10^{-4} will split into two equal parts at the junction with the equal sized parallel arms. The total flux in the circuit is thus 1.5×10^{-4} Wb. Hence

m.m.f. $= NI = 600I = 1.5 \times 10^{-4} \times 1.4 \times 10^6$

$I = 0.35$ A

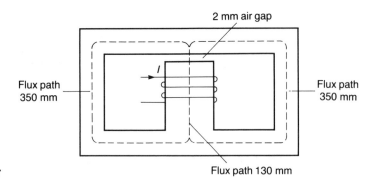

Fig. 4.13 Example 7

Example 7

The central limb of the magnetic circuit shown in Fig. 4.13 is wound with a coil of 600 turns of wire and has a cross-sectional area of 900 mm². The outer limbs of the circuit have a cross-sectional area of 500 mm². What is the current needed to produce a flux of 1.2 mWb in the central limb if the air gap in the central core has a length of 2 mm? The following data indicate the relationship between flux density and the magnetic field strength in the core material.

B (T)	1.0	1.1	1.2	1.3	1.4
H (A/m)	400	500	600	1000	1500

Answer

The flux density in the central limb will be

$$B = \frac{\Phi}{A} = \frac{1.2 \times 10^{-3}}{900 \times 10^{-6}} = 1.3 \text{ T}$$

From the data given this means a magnetic field strength H of 1000 A/m. Therefore the m.m.f. for the central limb is, using equation [5],

$$\text{m.m.f.} = HL = 1000 \times 0.13 = 130 \text{ A}$$

Because the two outer limbs are identical, the flux from the central limb must divide into two equal parts on passing out of the central limb. Thus the flux in an outer limb is 0.6×10^{-3} Wb. Hence the flux density in the outer limbs is

$$B = \frac{\Phi}{A} = \frac{0.6 \times 10^{-3}}{500 \times 10^{-6}} = 1.2 \text{ T}$$

From the data given this means a magnetic field strength H of 600 A/m. Therefore the m.m.f. required to provide this for an outer limb is

$$\text{m.m.f.} = HL = 600 \times 0.35 = 210 \text{ A}$$

The air gap will have the same flux passing through it as the central core and since it has the same cross-sectional area then it will have the same flux density of 1.3 T. The magnetic field strength for the air gap can be calculated using

$$B = \mu_r \mu_0 H$$

$$1.3 = 1 \times 4\pi \times 10^{-7} H$$

Hence H is 1.0×10^6 A/m. The m.m.f. required for the air gap is thus

$$\text{m.m.f.} = HL = 1.0 \times 10^6 \times 0.002 = 2000 \text{ A}$$

The total m.m.f. supplied by the coil will be the sum of the m.m.f. across the central limb plus that across the air gap plus that across the parallel arrangement of outer limbs (if it helps, solve the equivalent electrical circuit). Thus

$$\text{total m.m.f.} = 130 + 210 + 2000 = 2340 \text{ A}$$

Hence the current I required in the coil in order to supply this m.m.f. is given by

$$\text{m.m.f.} = NI = 600I = 2340$$

$$I = 3.9 \text{ A}$$

Magnetic leakage and fringing

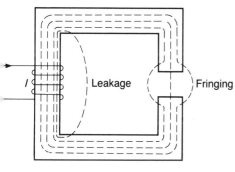

Fig. 4.14 Magnetic leakage and fringing

In the above discussions of magnetic circuits it has been assumed that all the flux is confined to the magnetic circuit and that none escapes to the surroundings. There is however invariably some *leakage* of flux. In addition, where there is a discontinuity such as an air gap in the circuit some of the flux by-passes it. The term used is *fringes*. Figure 4.14 illustrates both these effects.

The ratio of the total flux to the useful flux is called the *magnetic leakage coefficient*, the useful flux being that which has not leaked or fringed.

$$\text{Leakage coefficient} = \frac{\text{total flux}}{\text{useful flux}} \qquad [13]$$

The coefficient has typically values between about 1.05 and 1.4, the upper value indicating a very leaky magnetic circuit.

Example 8

An iron rod of length 50 cm and cross-sectional area 4 cm² is bent into a ring such that there is an air gap of 2 mm between the ends of the rod. The ring is wound with 400 turns of wire and a current of 1 A passes through it. If the relative permeability of the iron can be taken as 600 and the leakage coefficient is 1.2, what is the flux density in the air gap?

Answer

For the iron, the reluctance is given by equation [10] as

$$S = \frac{L}{\mu_r \mu_0 A} = \frac{0.50}{600 \times 4\pi \times 10^{-7} \times 4 \times 10^{-4}}$$

$$= 6.6 \times 10^6 \text{ A/Wb}$$

For the air gap, the reluctance is given by

$$S = \frac{L}{\mu_r \mu_0 A} = \frac{0.002}{1 \times 4\pi \times 10^{-7} \times 4 \times 10^{-4}}$$

$$= 4.0 \times 10^6 \text{ A/Wb}$$

The total reluctance is thus 10.6×10^6 A/Wb. The flux Φ produced for the circuit is given by m.m.f. $= \Phi S$ (equation [9]), hence

$$\text{total flux} = \frac{\text{m.m.f.}}{S} = \frac{NI}{S} = \frac{400 \times 1}{4.0 \times 10^6}$$

$$= 1.0 \times 10^{-4} \text{ Wb}$$

Because of leakage the useful flux in the air gap will be given by equation [13] as

$$\text{leakage coefficient} = \frac{\text{total flux}}{\text{useful flux}}$$

$$\text{useful flux} = \frac{1.0 \times 10^{-4}}{1.2}$$

$$= 8.3 \times 10^{-5} \text{ Wb}$$

The flux density B in the air gap is

$$B = \frac{\Phi}{A} = \frac{8.3 \times 10^{-5}}{4.0 \times 10^{-4}} = 0.21 \text{ T}$$

Magnetic screening

There are situations where it is necessary for an instrument to be shielded from magnetic fields. This can be achieved by surrounding it with a magnetic screen made of material which has a very low reluctance. Because for flux the lowest reluctance path is to go through the screen rather than the instrument space the flux passes mainly through the screen. It is like connecting a low-resistance resistor and a high-resistance resistor in parallel across a source of e.m.f.; most of the current will flow through the low-resistance resistor.

Inductance

When the current in a circuit changes, the magnetic flux originating from that current must change. This changes the flux linked by the circuit and hence an e.m.f. is induced which opposes the change producing it. The circuit is said to have *inductance*.

The induced e.m.f. is proportional to the rate of change of linked flux (see earlier in this chapter). The rate of change of flux will be proportional to the rate of change of the current dI/dt responsible for it. Thus

$$\text{e.m.f. is proportional to } -\frac{\mathrm{d}I}{\mathrm{d}t}$$

The constant of proportionality L is called the *inductance* of the circuit. Thus

$$\text{e.m.f.} = -\frac{L\,\mathrm{d}I}{\mathrm{d}t} \qquad [14]$$

The inductance of a circuit is said to be 1 henry (H) when the e.m.f. induced is 1 V as a result of the current changing by 1 A/s.

Example 9

What is the mean e.m.f. induced in a coil of inductance 200 mH when the current is increased from 1.0 A to 3.0 A in 0.05 s?

Answer

Using equation [14]

$$\text{e.m.f.} = -\frac{L\,\mathrm{d}I}{\mathrm{d}t} = -\frac{0.200 \times (3.0 - 1.0)}{0.05} = 8.0 \text{ V}$$

Flux linkage per ampere

The e.m.f. induced in a coil of N turns when the flux changes at the rate $\mathrm{d}\Phi/\mathrm{d}t$ is given by equation [2] as

$$\text{e.m.f.} = -\frac{N\mathrm{d}\Phi}{\mathrm{d}t}$$

This e.m.f. is related to the inductance L of the coil and the rate of change of current by equation [14]

$$\text{e.m.f.} = -\frac{L\,\mathrm{d}I}{\mathrm{d}t}$$

Hence

$$\frac{N\,\mathrm{d}\Phi}{\mathrm{d}t} = \frac{L\,\mathrm{d}I}{\mathrm{d}t}$$

$$L = \frac{N\,\mathrm{d}\Phi}{\mathrm{d}I} = \frac{\text{change in flux linkages}}{\text{change in current}}$$

For a coil having a constant reluctance, if the current changes from 0 to I and the flux changes from 0 to Φ, then

$$L = \frac{N\Phi}{I} \qquad [15]$$

We can consider the inductance to be the flux linkages produced per ampere.

Example 10

What is the flux produced in a coil of 400 turns by a current of 5.0 A if it has a constant inductance of 10 mH?

Answer

Using equation [15]

$$L = \frac{N\Phi}{I}$$

$$\Phi = \frac{LI}{N} = \frac{0.010 \times 5.0}{400} = 1.25 \times 10^{-4} \text{ Wb}$$

Inductance of a solenoid

For a coil of N turns and length l carrying a current I, if l is taken as being the length of the flux path then (equation [5])

$$\text{magnetic field strength } H = \frac{IN}{l}$$

If the core of the coil has a constant relative permeability μ_r, then

$$\text{flux density } B = \mu_r\mu_0 H = \mu_r\mu_0\frac{IN}{l}$$

Hence the flux Φ is

$$\Phi = BA = \mu_r\mu_0\frac{INA}{l}$$

where A is the cross-sectional area of the coil. The inductance L of the coil is thus given by

$$L = \frac{N\Phi}{I} = \frac{\mu_r\mu_0 N^2 A}{l} \tag{16}$$

or, since the reluctance $S = l/\mu_r\mu_0 A$, then

$$L = \frac{N^2}{S} \tag{17}$$

The above derivation has assumed a constant relative permeability; this will not be the case if the flux density varies with the magnetic field strength in the way described in Fig. 4.4. A consequence of such a variation is that the inductance depends on the current. Typically, for magnetic materials such as iron the inductance will vary with current in the manner described by Fig. 4.15.

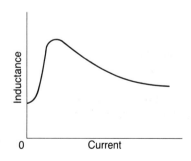

Fig. 4.15 Variation of inductance with current for an iron-cored coil

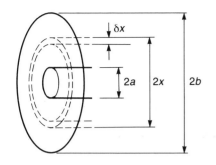

↓δx

2a 2x 2b

Fig. 4.16 Coaxial cable

Inductance of coaxial cable

Example 11

What is the inductance of an air-cored coil of wire if it has 300 turns, a cross-sectional area of 400 mm² and a length of 200 mm?

Answer

Using equation [16] and assuming that the length of the flux path is the length of the coil,

$$L = \frac{\mu_r\mu_0 N^2 A}{l} = \frac{1 \times 4\pi \times 10^{-7} \times 300^2 \times 400 \times 10^{-6}}{0.200}$$

$$= 2.3 \times 10^{-4} \text{ H}$$

Consider a coaxial cable with an inner conductor of radius a and an outer conductor with an internal radius b, as in Fig. 4.16. There is a current I in one direction in the core and a current I in the opposite direction in the outer conductor. These oppositely directed currents mean that outside the cable the magnetic field is zero. The magnetic field strength H at some radial distance x, where x has a value between a and b, is given by equation [5] as

$$H = \frac{NI}{l}$$

but N equals 1 and l the length of the flux path at radius x is $2\pi x$. Hence

$$H = \frac{I}{2\pi x}$$

The flux density B at radius x is given by equation [6] as

$$B = \mu_r\mu_0 H = \frac{\mu_r\mu_0 I}{2\pi x}$$

The flux $\delta\Phi$ within an element of width δx and a length of conductor of 1 m, is thus

$$\delta\Phi = BA = B\delta x = \frac{\mu_r\mu_0 I\delta x}{2\pi x}$$

The total flux Φ linked per metre between the core and the outer sheath is thus

$$\Phi = \int_a^b \frac{\mu_r\mu_0 I}{2\pi x}\, dx = \frac{\mu_r\mu_0 I}{2\pi} \int_a^b \frac{1}{x}\, dx$$

$$= \frac{\mu_r\mu_0 I}{2\pi}(\ln b - \ln a) = \frac{\mu_r\mu_0 I}{2\pi} \ln \frac{b}{a}$$

Hence since the inductance is Φ/I, then the inductance per metre L is

$$L = \frac{\mu_r \mu_0}{2\pi} \ln \frac{b}{a} \qquad [18]$$

The above discussion neglects any consideration of the inductance that the inner conductor would have due to flux linkage within itself as a result of the current through it. This is most pronounced with low-frequency or d.c. current because only then is the current uniformly distributed over the cross-section of the conductor. At high frequencies the current tends to become concentrated at the conductor surfaces. Under low-frequency or d.c. conditions a correction term is added to the above equation of $(\mu_r \mu_0/8\pi)$.

Example 12

What is the inductance per metre length of coaxial cable if it has an inner core of radius 1.0 mm and an outer conductor of internal radius 4.0 mm? The relative permeability of the medium between the conductors can be taken as 1 and any inductance due to internal flux linkages in the core neglected.

Answer

Using equation [18] for the inductance per metre,

$$L = \frac{\mu_r \mu_0}{2\pi} \ln \frac{b}{a} = \frac{1 \times 4\pi \times 10^{-7}}{2\pi} \ln \frac{4.0}{1.0} = 2.8 \times 10^{-7} \text{ H/m}$$

Example 13

What is the inductance per metre length of the coaxial cable specified in example 12 if internal flux linkages in the core are not neglected?

Answer

The inductance per metre due to internal flux linkages in the core is

$$L = \frac{\mu_r \mu_0}{8\pi} = \frac{1 \times 4\pi \times 10^{-7}}{8\pi} = 0.5 \times 10^{-7} \text{ H/m}$$

Hence the total inductance per metre for the coaxial cable is 3.3×10^{-7} H/m.

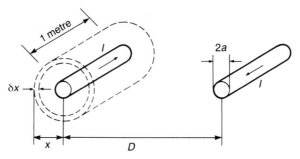

Fig. 4.17 Parallel transmission lines

Inductance of parallel transmission lines

Consider two parallel conductors, each of radius a, with the distance between their centres D (Fig. 4.17). D is assumed to be considerably greater than a. To determine the flux between the conductors we can consider the flux due to each conductor in turn and add the two. We are only concerned with the flux between the wires since that is the flux linked by the circuit formed by the wires. The flux due to one wire which extends beyond the other wire does not contribute to the self inductance.

The magnetic field strength H_x at some distance x from one wire, since there is but one turn and the flux path at this distance is $2\pi x$, is,

$$H_x = \frac{NI}{l} = \frac{I}{2\pi x}$$

The flux density B_x at this distance is thus

$$B_x = \mu_r\mu_0 H = \frac{\mu_r\mu_0 I}{2\pi x}$$

For a 1 m length of wire, the flux passing through an element of width δx will be

$$\Phi = BA = B\delta x = \frac{\mu_r\mu_0 I\delta x}{2\pi x}$$

The total flux due to this one wire between a distance a and D is thus

$$\text{total flux due to one wire} = \int_a^D \frac{\mu_r\mu_0 I}{2\pi x}\,dx$$

$$= \frac{\mu_r\mu_0 I}{2\pi}\ln\frac{D}{a}$$

The flux due to the other wire between the wires will have the same value and be in the same direction. Hence the total flux linked is

$$\text{total flux} = \frac{\mu_r\mu_0 I}{\pi}\ln\frac{D}{a}$$

The inductance L per metre length of the wires is given by $L = N\Phi/I$, hence since N is 1,

$$L = \frac{\mu_r\mu_0}{\pi}\ln\frac{D}{a} \qquad\qquad [19]$$

The above equation takes no account of internal flux linkages within each wire. This will, for d.c. and low frequencies, contribute for each wire an inductance of $(\mu_r\mu_0/8\pi)$ per metre and so for the two wires $2(\mu_r\mu_0/8\pi)$ should be added to the above equation.

Example 14

Ignoring internal linkages, what is the inductance per metre length of a single-phase power line which consists of two conductors each of radius 10 mm and 900 mm apart in air?

Answer

Using equation [19]

$$L = \frac{\mu_r\mu_0}{\pi} \ln \frac{D}{a} = \frac{1 \times 4\pi \times 10^{-7}}{\pi} \ln \frac{900}{10} = 1.8 \times 10^{-6} \text{ H/m}$$

Example 15

What would be the inductance per metre length of the power line specified in example 14 if internal linkages were not neglected?

Answer

The inductance per metre for the two wires due to internal linkages is

$$L = \frac{2\mu_r\mu_0}{8\pi} = \frac{2 \times 1 \times 4\pi \times 10^{-7}}{8\pi} = 2 \times 10^{-7} \text{ H/m}$$

Hence the total inductance is 2.0×10^{-6} H/m.

Mutal inductance

When the current in a coil changes, the magnetic flux of that coil changes. If this flux links another coil then an e.m.f. will be induced in it. In such a situation the coils are said to be *magnetically coupled* and the effect is called *mutual inductance*.

The flux linked by the secondary coil depends on the current I_1 in the primary coil. The e.m.f. induced in the secondary e_2 is proportional to the rate of change of flux linked by it. Hence the e.m.f. induced in the secondary coil must be proportional to the rate of change of current in the primary coil.

$$\text{induced e.m.f. } e_2 \text{ is proportional to } - \frac{\mathrm{d}I_1}{\mathrm{d}t}$$

$$e_2 = - M \frac{\mathrm{d}I_1}{\mathrm{d}t} \tag{20}$$

M is a constant for the arrangement of coils concerned and is called the *mutual inductance*. The unit of mutual inductance is the henry (H). The minus sign is because the induced e.m.f. tends to circulate a current in the secondary coil in such a direction as to oppose the change in flux produced by the current change in the primary coil.

If the current in the primary coil changes by δI_1 in a time δt and produces a flux change of $\delta\Phi$ in the secondary coil, number of turns N_2, in that time then the induced e.m.f. in the secondary will be, by equation [2],

$$e_2 = -N_2 \frac{\delta\Phi}{\delta t}$$

But we can also write

$$e_2 = -M \frac{\delta I_1}{\delta t}$$

Hence

$$M \frac{\delta I_1}{\delta t} = N_2 \frac{\delta\Phi}{\delta t}$$

$$M = N_2 \frac{\delta\Phi}{\delta I_1} \qquad [21]$$

$N_2 \delta\Phi$ is the change of flux linkages with the secondary coil. Hence

$$M = \frac{\text{change of flux linkages with the secondary coil}}{\text{change in current in the primary coil}} \qquad [22]$$

If the relative permeability of the magnetic circuit remains constant, this also means a constant reluctance, then if Φ is the linked flux when the current in the primary is I_1,

$$M = \frac{N_2 \Phi}{I_1}$$

The mutual inductance between two circuits is the same whichever one produces the flux change. Thus we could have derived the above relationships for the e.m.f. induced in the primary as a result of a current change in the secondary. Thus

$$M = N_1 \frac{\delta\Phi}{\delta I_2}$$

Example 16

If the mutual inductance of a pair of air-cored coils is 200 µH, what will be the e.m.f. induced in the secondary coil when the current in the primary coil changes at the rate of 2000 A/s?

Answer

Using equation [20]

$$\text{induced e.m.f.} = M \frac{dI_1}{dt} = 200 \times 10^{-6} \times 2000 = 0.4 \text{ V}$$

Coefficient of magnetic coupling

The *magnetic coupling coefficient k* is used to describe the degree of magnetic coupling that occurs between circuits.

$$k = \frac{\text{flux linking the two circuits}}{\text{total flux produced}} \quad [23]$$

If there is no magnetic coupling then k is zero. If the magnetic coupling is perfect and all the flux produced in the primary links with the secondary then k is 1. Where k is low then circuits are said to be loosely coupled, where k is high, tightly coupled.

Consider a primary coil of inductance L_1 which is coupled with a secondary coil of inductance L_2. If each has a core of constant relative permeability, then a current of I_1 in the primary coil will produce a flux of Φ_1 through its turns. Using the equation [15] developed earlier in this chapter,

$$L_1 = \frac{N_1 \Phi_1}{I_1}$$

and thus

$$\Phi_1 = \frac{L_1 I_1}{N_1}$$

The flux that links the secondary coil will be $k\Phi_1$. Thus if the coils have a mutual inductance M, using the equation [21] developed above

$$M = \frac{N_2 k \Phi_1}{I_1}$$

$$= \frac{k N_2 L_1}{N_1}$$

Similarly, if coil 2 is regarded as the primary coil and coil 1 as the secondary

$$M = \frac{k N_1 L_2}{N_2}$$

Thus

$$\frac{N_2}{N_1} = \frac{M}{k L_1} = \frac{k L_2}{M}$$

$$k = \frac{M}{\sqrt{(L_1 L_2)}} \quad [24]$$

Example 17

What is the magnetic coupling coefficient of a pair of coils if they have inductances of 200 mH and 400 mH and the mutual inductance is 50 mH?

Answer

Using equation [24]

$$k = \frac{M}{\sqrt{(L_1 L_2)}} = \frac{0.050}{\sqrt{(0.200 \times 0.400)}} = 0.18$$

Dot notation

It is not clear from circuit diagrams of coupled circuits whether the e.m.f.s due to currents in the coils are in the same direction or not. To overcome this, dot notation is used. Figure 4.18 shows the notation. A dot is put near one end of a coil and then another dot is placed at the end of the magnetically coupled coil which has the same instantaneous polarity as the dotted end of the first marked coil.

With this notation, if a current enters the end of the coil marked with a dot then the induced e.m.f. in the coupled coil will make that end of the coupled coil marked with a dot positive. If a current leaves the end of a coil marked with a dot then the induced e.m.f. in the end of the coupled coil marked with a dot will be negative.

Coupled circuits

Consider a coupled pair of coils with a mutual inductance M, as in Fig. 4.18. If the primary coil is supplied with a sinusoidal alternating current then the current I_1 at any instant is given by

$$I_1 = I_{1m}\sin \omega t$$

where I_{1m} is the maximum value of the current and $\omega = 2\pi f$, where f is the frequency. The e.m.f. e_2 induced in the secondary coil is

$$e_2 = -M \frac{dI_1}{dt} = -M \frac{d(I_{1m}\sin \omega t)}{dt}$$

$$e_2 = -MI_{1m}\omega \cos \omega t = -MI_{1m}\omega \sin(\omega t = 90°)$$

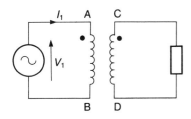

Fig. 4.18 Example of dot notation

This e.m.f. is 90° out of phase with the current, and applied voltage in the primary circuit. We can represent this e.m.f. as $j\omega MI_1$ and so produce an equivalent secondary which consists of a voltage generator of size $j\omega MI_1$ in series with the coil. Similarly for the primary coil, we can produce an equivalent circuit which consists of a voltage generator of size $j\omega MI_2$ in series with the coil.

The direction of these e.m.f.s is indicated by the dot

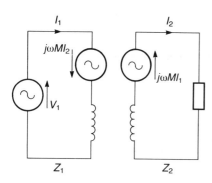

Fig. 4.19 Equivalent circuits for coupled coils

notation. Thus for the dot notation used in Fig. 4.18, Current I_1 enters point A, the dot end of the primary coil, and so the induced e.m.f. $j\omega MI_1$ in the dot end of the secondary coil, point C, will be positive. Current I_2 leaves the dot end of the secondary coil, C, and so the dot end of the primary coil, A, will be negative. Hence the equivalent circuits for the two coils are as shown in Fig. 4.19.

For the primary circuit we can thus write

$$V_1 = I_1 Z_1 - j\omega MI_2$$

and for the secondary circuit

$$0 = -j\omega MI_1 + I_2 Z_2$$

where Z_1 is self-impedance of the primary circuit and Z_2 the self-impedance of the secondary circuit. Hence from these two equations

$$I_2 = \frac{j\omega MI_1}{Z_2} \qquad [25]$$

$$V_1 = I_1(Z_1 + \omega^2 M^2/Z_2) \qquad [26]$$

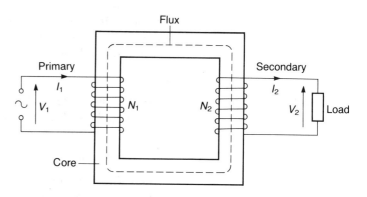

Fig. 4.20 The basic form of a transformer

Transformers

The principle behind the operation of a transformer is mutual inductance between two coils. Figure 4.20 shows the basic form of a transformer. An alternating voltage is applied to the primary coil. This produces an alternating current in the coil and so an alternating flux in the core. The secondary coil is wound on the same core and the flux produced by the primary coil links its turns. If there are no flux losses from the magnetic circuit, the flux linking each turn of the secondary coil is the same as the flux linking each turn of the primary coil. Thus the e.m.f. induced per turn of either the secondary or primary coils by the changing flux is the same. Hence

$$\frac{\text{induced e.m.f. in primary}}{\text{induced e.m.f. in secondary}} = \frac{N_1}{N_2}$$

When the secondary coil is on open circuit, i.e., there is no load in its circuit, then the voltage between the terminals of the coil is the same as the induced e.m.f. If there is no load then there is no current and so no energy is taken from the secondary coil. This means, for an ideal transformer, that no energy is taken from the primary coil and so there will be no current in the primary circuit. This can only be the case if the induced e.m.f. is equal to and opposing the input voltage. Thus

$$\frac{\text{supply voltage } V_1}{\text{secondary voltage } V_2} = \frac{N_1}{N_2} \qquad [27]$$

If V_2 is less than V_1 the transformer is said to have a step-down voltage ratio, if V_2 is more than V_1 a step-up voltage ratio.

For an ideal transformer there will be no power loss and thus the volt-amperes supplied to the primary = volt-amperes supplied to the load

$$I_1 V_1 = I_2 I_2$$

Thus

$$\frac{V_1}{V_2} = \frac{I_2}{I_1} = \frac{N_1}{N_2}$$

This yields the equation

$$I_1 N_1 = I_2 N_2 \qquad [28]$$

Thus, the number of ampere-turns on the secondary winding equals the number of ampere-turns in the primary winding.

In specifying a transformer a number of parameters are used; the volt-amperes supplied to the primary, the turns ratio, and the primary to secondary voltage ratio. Thus, for example, a transformer might be rated as 50 kVA, 1000/10 V.

Example 18

What will be the secondary voltage produced if a single-phase transformer, with 200 primary turns and 50 secondary turns, has an a.c. input of 240 V (r.m.s.)?

Answer

Using the equation [27] developed above

$$\frac{V_1}{V_2} = \frac{N_1}{N_2}$$

$$V_2 = \frac{240 \times 50}{200} = 80 \text{ V}$$

The e.m.f. equation for a transformer

The alternating current in the primary coil of a transformer gives rise to alternating flux. If the current is sinusoidal of frequency f then since the flux Φ will be proportional to the current the flux varies with time according to an equation of the form

$$\Phi = \Phi_m \sin 2\pi ft$$

where Φ_m is the maximum value of the flux, Φ is the flux at time t. The induced e.m.f. per turn of both primary and secondary coils is thus

$$\text{e.m.f. per turn} = -\frac{d\Phi}{dt} = -\frac{d(\Phi_m \sin 2\pi ft)}{dt}$$

$$= -2\pi f\Phi_m \cos 2\pi ft$$

The maximum value of the e.m.f. will be when the cosine term has its maximum value of 1. Hence

$$\text{max. e.m.f. per turn} = 2\pi f\Phi_m$$

The r.m.s. value of the e.m.f. per turn is (max. e.m.f.)/$\sqrt{2}$, hence

$$\text{r.m.s. e.m.f. per turn} = \frac{2\pi f\Phi_m}{\sqrt{2}}$$

$$= 4.44\, f\Phi_m \qquad\qquad [29]$$

Example 19

What is the maximum value of the flux in the core of an ideal transformer if the primary has 500 turns and the input is 240 V at 50 Hz?

Answer

The e.m.f. per turn is 240/500 = 0.48 V. Hence using the equation [29] developed above

$$\text{r.m.s. e.m.f. per turn} = 4.44\, f\Phi_m$$

$$\Phi_m = \frac{0.48}{4.44 \times 50} = 2.2 \times 10^{-3}\ \text{Wb}$$

Flux leakage

When there is no load in the secondary of a transformer the current in the primary I_0 can be considered to be composed of two parts. One part, called the magnetising component I_{mag}, is responsible for producing the flux. This component is thus in phase with the flux. The other part of the primary current is responsible for supplying the hysteresis and eddy current losses in the core and the I^2R loss in the primary. This component I_c is in phase with the applied **primary voltage**

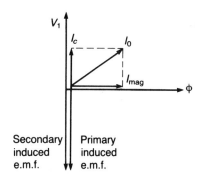

Fig. 4.21 Phasor diagram for a transformer on no-load

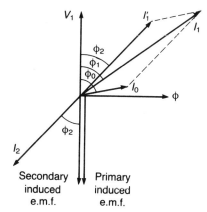

Fig. 4.22 Phasor diagram for a loaded transformer with negligible voltage drop in the windings

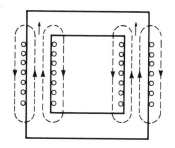

Fig. 4.23 Paths of leakage flux

Leakage reactances

V_1 and so 90° out of phase with the magnetising component. Figure 4.21 shows the phasor diagram for this no-load condition.

$$\text{No load current } I_0 = \sqrt{(I_c^2 + I_{mag}^2)} \qquad [30]$$

With the transformer operating on no-load the flux passing through the air surrounding the core will be negligible compared to that passing through the core. This is because the relative permeability of the core is considerably larger than that of air, possibly about a thousand times greater. Thus the reluctance of the core path for the flux is considerably smaller than that of the air path.

When there is a load in the secondary then there is a current in the secondary I_2. This will give rise to flux and hence a current in the primary. The primary current has thus three components, two of them as before on open circuit and one I'_1 to neutralise the demagnetising effect of the secondary current. This component will be in the opposite direction and equal to the secondary current I_2. The angle ϕ_2 between the secondary current I_2 and the secondary voltage V_2 phasors will depend on the power factor of the load. Figure 4.22 shows a phasor diagram for a loaded transformer when there is negligible voltage drop in the windings.

With a load in the secondary, the magnetising current will generally not be in phase with the component to neutralise the demagnetising effect of the secondary current. Thus at some instant the magnetising current will be zero and the secondary current and demagnetising current not. When this occurs if the direction of the flux is upwards in the primary core limb of the transformer, see Fig. 4.23, then the direction of the flux in the secondary limb will also be upwards. The return path for each of these fluxes has to be through the air and so is referred to as *leakage flux*. This flux is responsible for inducing an e.m.f. in the coil to which it is linked.

The amount of leakage flux depends on the load current, whereas the useful part of the flux due to the current I_{mag} is virtually independent of the load.

As discussed above, the leakage flux in the primary is proportional to the secondary current. The effect of this is to induce an e.m.f. in the primary windings. This is equivalent to having an inductive reactance connected in series with a transformer having no leakage. Likewise the leakage flux in the secondary is equivalent to having an inductive reactance

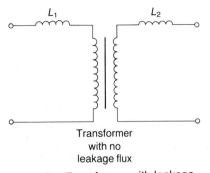

Transformer
with no
leakage flux

Fig. 4.24 Transformer with leakage reactances

connected in series with a transformer having no leakage. The resulting equivalent circuit is as shown in Fig. 4.24.

An equivalent circuit for a transformer which also allows for the resistances of the transformer windings is shown in Fig. 4.25(a). R_1 represents the resistance of the primary windings, R_2 that of the secondary. This circuit can be simplified if R'_2 is replaced by an additional resistance R$'_2$ in the primary circuit such that the power developed in this resistance when carrying the primary current is equal to that which would have been developed in R_2 due to the secondary current (Fig. 4.25(b)). Thus

$$I_1^2 R'_2 = I_2^2 R_2$$

$$R'_2 = R_2(I_2/I_1)^2 \approx R_2(N_1/N_2)^2$$

The total resistance R_e which is equivalent to the resistances of both windings is thus

$$R_e = R_1 + R_2(N_1/N_2)^2 \qquad [31]$$

In a similar manner we can replace the secondary leakage reactance X_2 by an equivalent reactance X'_2 in the primary circuit. The inductance of a coil is proportional to the square of the number of turns. Thus

$$X'_2 = X_2(N_1/N_2)^2$$

(a)

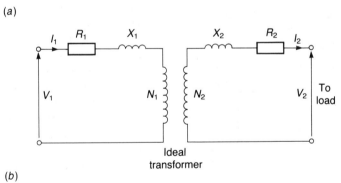

(b)

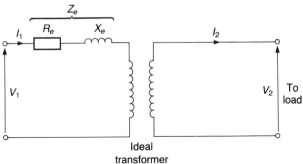

Fig. 4.25 (a) An equivalent circuit for a transformer (b) a simplified version of the circuit

The total reactance X_e which is equivalent to the primary and secondary leakage reactances is thus

$$X_e = X_1 + X_2(N_1/N_2)^2 \qquad [32]$$

The equivalent impedance Z_e in the primary circuit for the primary and secondary windings is thus

$$Z_e = \sqrt{(R_e^2 + X_e^2)} \qquad [33]$$

A more complete equivalent circuit for a transformer is given in Fig. 4.26. In addition to accounting for the leakage reactances, account is also taken of the no-load current I_0 and its two components, the magnetising current I_{mag} and the core loss current I_c. Since the value of the no-load current is generally less than about 5% of the full-load primary current the circuit branch for it is often neglected and just the circuit shown in Fig. 4.25 used.

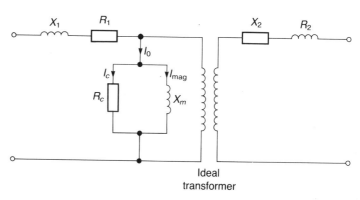

Fig. 4.26 The complete equivalent circuit for a transformer

Example 20

A transformer has 500 primary turns and 100 secondary turns. The primary resistance is 0.40 Ω and the secondary 0.02 Ω. The primary leakage reactance is 1.20 Ω and the secondary 0.03 Ω. What is the equivalent impedance which could be placed in the primary circuit with an ideal transformer?

Answer

Using the equations [31], [32] and [33] developed above

$$\begin{aligned}
R_e &= R_1 + R_2(N_1/N_2)^2 \\
&= 0.40 + 0.02(500/100)^2 \\
&= 0.90 \ \Omega
\end{aligned}$$

$$\begin{aligned}
X_e &= X_1 + X_2(N_1/N_2)^2 \\
&= 1.20 + 0.03(500/100)^2 \\
&= 1.95 \ \Omega
\end{aligned}$$

$$\begin{aligned}
Z_e &= \sqrt{(R_e^2 + X_e^2)} \\
&= \sqrt{(0.90^2 + 1.95^2)} \\
&= 2.15 \ \Omega
\end{aligned}$$

Minimising leakage flux

Leakage flux can be considerably reduced by:

1 making the central opening in the transformer core tall and narrow;
2 sandwiching the primary and secondary windings, as in Fig. 4.27;
3 using a shell form of construction, as in Fig. 4.28.

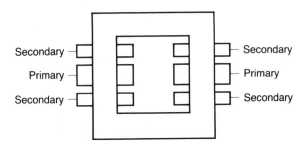

Fig. 4.27 Sandwiched transformer windings

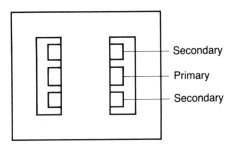

Fig. 4.28 Shell form of transformer

Power losses in a transformer

The power losses with a transformer can be considered to be classified as *iron losses* (or *core losses*) and *copper losses*. Iron losses are the power dissipated in the magnetic material used for the transformer core. The iron loss is essentially constant and independent of the currents in the coils. The iron loss consists of two forms of losses, eddy current losses and hysteresis losses. These are both discussed later in this chapter.

Copper losses are the I^2R losses in the copper conductors of the primary and secondary windings which result from currents flowing through them.

$$\text{Copper loss} = I_1^2 R_1 + I_2^2 R_2 \qquad [34]$$

where R_1 is the resistance of the primary coil and I_1 the current through it, R_2 is the resistance of the secondary coil and I_2 the current through it.

Eddy current losses

The term *eddy currents* is given to the currents induced in a conductor as a result of flux changes within that conductor.

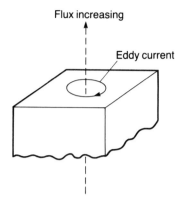

Flux increasing

Eddy current

Fig. 4.29 Eddy currents

Thus for a magnetic circuit, as for example a transformer, when the flux in the core changes then an e.m.f. is induced which gives rise to eddy currents in the core material (Fig. 4.29). These induced currents circulate in the core in such a direction as to produce flux which opposes the change responsible for their production. Since the core material is generally a good conductor of electricity the size of the induced currents can be large. The power loss, the I^2R loss, associated with these currents can thus be quite large.

The size of the eddy currents can be reduced by using a laminated core instead of a solid one. The laminations are thin sheets insulated from each other. The eddy currents then become confined to a lamination. This reduces the size of the currents because the thinner the sheet, the smaller its cross-sectional area, and so the larger the resistance (resistance = resistivity × length/area). Hence

resistance is proportional to $1/t$

where t is sheet thickness.

The factor determining the size of the induced e.m.f. and hence eddy current is the rate at which the flux is changing. Thus since the rate at which the flux is changing is proportional to the frequency f of the alternating flux and the flux is proportional to the flux density

induced e.m.f. is proportional to fB_m

where B_m is the maximum value of the alternating flux. Hence

I^2R loss per lamination is proportional to
$$\text{(induced e.m.f.)}^2/\text{resistance}$$

I^2R loss per lamination is proportional to $(fB_m)^2t$

The smaller the thickness t the greater the number of laminations, hence the total I^2R loss for the core is

total I^2R loss is proportional to $f^2B_m{}^2t^2$

This relationship is normally written as

$$P_e = k_e B_m{}^2 f^2 t^2 \qquad [35]$$

where P_e is the eddy current loss per cubic metre of core and k_e is a constant.

Example 21

A transformer has an eddy current loss of 80 W when operating at a frequency of 50 Hz. What will be the eddy current loss at a frequency of 60 Hz if the flux density remains unchanged?

Answer

As discussed above, the eddy current loss is proportional to the square of the frequency. Thus the loss P at 60 Hz will be

$$\frac{P}{80} = \frac{60^2}{50^2}$$

$$P = 115 \text{ W}$$

Hysteresis losses

See Chapter 5 for a discussion of hysteresis. When the current through a coil is reversed the state of magnetisation of the core is reversed. Each time this occurs energy is expended, this being known as the *hysteresis loss*. The hysteresis power loss, P_h, is proportional to the volume v of material for which the magnetisation is being continually reversed.

P_h is proportional to volume v

The power loss for each cycle of reversal depends on the maximum flux density B_m produced in the material,

P_h is proportional to $B_m{}^n$

where n is a number known as the *Steinmetz index*. It usually has values in the range 1.6 to 2.2. The power loss also depends on the frequency f with which the magnetisation is reversed, doubling the frequency doubles the power loss.

P_h is proportional to f

The above factors can be combined in the relationship

$$P_h = k_h v f B_m{}^n \qquad [36]$$

where k_h is a constant for a given core and range of flux density.

Example 22

The hysteresis loss of a transformer is 300 W when it is used with a 240 V, 50 Hz supply. What will be the hysteresis loss when it is used with a 110 V, 60 Hz supply? Take the Steinmetz index to be 1.6.

Answer

Using the equation [36] developed above

$$P_h = k_h v f B_m{}^n$$

and then using equation [29]

r.m.s. e.m.f. per turn $= 4.44\, f\Phi_m$

The maximum flux density B_m is proportional to the maximum flux developed Φ_m and hence is proportional to (r.m.s. e.m.f. per turn)/f. Thus, since the number of turns is constant

B_m is proportional to (applied voltage V)/f

Hence

P_h is proportional to $f(V/f)^n$

Hence

$$\frac{P_h \text{ under new conditions}}{P_h \text{ initially}} = \frac{60(110/60)^{1.6}}{50(240/50)^{1.6}}$$

Hence the hysteresis loss under the new conditions is 77 W.

Iron losses

The iron losses P_c in a magnetic circuit are the sum of the hysteresis P_h and eddy current P_e losses. Thus, using the equations [35] and [36] developed above,

$$P_c = P_h + P_e$$
$$= k_h v f B_m{}^n + k_e B_m{}^2 f^2 t^2$$

If, for a particular inductor or transformer with v and t constant, the flux density is constant then

$$P_c = k_1 f + k_2 f^2$$

$$\frac{P_c}{f} = k_1 + k_2 f$$

where $k_1 = k_h v B_m{}^n$ and $k_2 = k_e B_m{}^2 t^2$.

A graph of P_c/f plotted against f is a straight line with an intercept with the P_c/f axis of k_1 and a slope of k_2. Thus a measurement of the iron loss at different frequencies enables the separate contributions of the hysteresis and eddy current losses to be determined.

Example 23

Following measurements of the iron loss of a transformer at different frequencies, a graph is plotted of (iron loss/frequency) against frequency. The graph is a straight line which intercepts the (iron loss/frequency) axis at a value of 0.6 W/Hz and has a slope of 0.040 W/Hz^2. What will be the eddy current and hysteresis losses at 50 Hz?

Answer

The equation of the straight line graph is

$$\frac{P_c}{f} = k_1 + k_2 f$$

Hence $k_1 = 0.6$ W/Hz. But the hysteresis loss P_h is given by

$$P_h = k_1 f = 0.6 \times 50 = 30 \text{ W}$$

The slope of the graph is k_2. Hence $k_2 = 0.040$ W/Hz^2. Thus the eddy current loss P_e is given by

$$P_e = k_2 f^2 = 0.040 \times 50^2 = 100 \text{ W}$$

Transformer regulation

Increasing the load in the secondary circuit of a transformer increases the losses and hence there is a difference between the secondary voltage at no load and at full load. This effect is described by

per-unit voltage regulation

$$= \frac{\text{no load voltage} - \text{full load voltage}}{\text{no load voltage}} \qquad [37]$$

Transformer efficiency

The per-unit efficiency η of a transformer is given by

$$\eta = \frac{\text{output power}}{\text{input power}} \qquad [38]$$

Since input power = output power + losses

$$\eta = \frac{\text{output power}}{\text{output power} + \text{losses}} \qquad [39]$$

$$\eta = \frac{\text{input power} - \text{losses}}{\text{input power}}$$

Example 24

What is the efficiency of a transformer which delivers 3 kW of power to a load and under these conditions has an iron loss of 50 W and a copper loss of 70 W?

Answer

Using the equation [39] given above

$$\text{efficiency} = \frac{\text{output power}}{\text{output power} + \text{losses}}$$

$$= \frac{3000}{3000 + 120} = 0.96 \text{ or } 96\%$$

Example 25

What is the efficiency of a 5 kVA, single-phase transformer, operating at half load if it has an iron loss of 80 W and a full-load copper loss of 120 W and the power factor of the load is 0.8?

Answer

The copper loss varies as the square of the current, therefore at half load the copper loss will be $120 \times (0.5)^2 = 30$ W. Therefore the total loss at half load is $(30 + 80) = 110$ W.

At 0.8 power factor the full load output is $5 \times 0.8 = 4$ kW. At half load the output is thus 2 kW. Hence the efficiency is, using equation [39],

$$\text{efficiency} = \frac{\text{output}}{\text{output} + \text{losses}} = \frac{2000}{2000 + 110} = 0.95 \text{ or } 95\%$$

Maximum efficiency

If R_{e2} is the equivalent resistance of the primary and secondary windings referred to the secondary circuit then the total copper loss is $I_2^2 R_e^2$ (see the discussion earlier in this chapter of leakage reactance). Then if P_c is the iron (or core) loss,

$$\text{efficiency} = \frac{\text{output power}}{\text{input power}}$$

$$= \frac{V_2 I_2 \cos \phi_2}{V_2 I_2 \cos \phi_2 + I_2^2 R_{e2} + P_c}$$

$$= \frac{V_2 \cos \phi_2}{V_1 \cos \phi_2 + I_2 R_{e2} + (P_c/I_2)}$$

The maximum efficiency is obtained by differentiating this equation with respect to I_2. Since V_2 and $\cos \phi_2$ are constant for a particular load this maximum efficiency is when the denominator has a minimum value, i.e.,

$$\frac{d(V_2 \cos \phi_2 + I_2 R_{e2} + P_c/I_2)}{dI_2} = 0$$

This is when

$$R_{e2} - P_c/I_2^2 = 0$$

and so requires

$$I_2^2 R_{e2} = P_c \qquad [40]$$

Thus maximum efficiency is achieved when the copper losses equal the iron losses.

Example 26

A 200 kVA transformer has an iron loss of 1.2 kW and a full-load copper loss of 2.8 kW. At what output and what efficiency will the efficiency of the transformer be a maximum? The power factor of the load may be taken as 0.8.

Answer

If n is the fraction of the full-load current at which the efficiency is a maximum, then, since the loss is proportional to I^2, the copper loss under these conditions will be

copper loss = $n^2 \times 2.8$ kW

At maximum efficiency the copper loss equals the iron loss. Thus

$n^2 \times 2.8 = 1.2$

$n = 0.65$

Therefore the output at maximum efficiency $= 0.65 \times 200 = 130$ kVA. The output power is thus $130 \times 0.8 = 104$ kW and the efficiency is

$$\text{maximum efficiency} = \frac{104}{104 + 2 \times 1.2} = 0.98 \text{ or } 98\%$$

Impedance matching

One use of a transformer is to change the apparent value of an impedance. For an ideal transformer

$$\frac{V_1}{V_2} = \frac{N_1}{N_2}$$

$$V_1 = \frac{N_1 V_2}{N_2}$$

where N_1 and N_2 are the primary and secondary turns, V_1 and V_2 the primary and secondary voltages. Also

$$\frac{I_1}{I_2} = \frac{N_2}{N_1}$$

$$I_1 = \frac{N_2 I_2}{N_1}$$

Hence

$$\frac{V_1}{I_1} = \left(\frac{N_1}{N_2}\right)^2 \frac{V_2}{I_2}$$

But V_1/I_1 is the impedance Z_1 into which the primary voltage source operates. V_2/I_2 is the impedance Z_2 of the load circuit. Thus

$$Z_1 = \left(\frac{N_1}{N_2}\right)^2 Z_2 \qquad [41]$$

Thus positioning a transformer between the source and the load alters the apparent impedance of the load.

According to the *maximum power transfer theorem* the

maximum power is transferred from one circuit to another when the impedance of the load circuit is equal to the impedance of the supply circuit. A transformer is frequently used to achieve this and is then said to be *impedance matching*.

Example 27

What should be the turns ratio of a transformer which is to be used to match an amplifier with an output impedance of 40 Ω with a circuit of input impedance 600 Ω and so give maximum power transfer?

Answer

Using the equation [41] developed above

$$Z_1 = \left(\frac{N_1}{N_2}\right)^2 Z_2$$

$$40 = \left(\frac{N_1}{N_2}\right)^2 600$$

$$\frac{N_1}{N_2} = 0.26$$

Force on current-carrying conductor

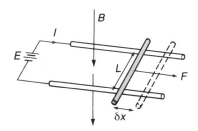

Fig. 4.30   Force on a current-carrying conductor

Consider the situation described by Fig. 4.30. A source of e.m.f. E gives rise to a current I through a conductor. The conductor rests on parallel rails a distance of L apart and there is a flux of density B at right-angles to the plane of the rails and conductor. The effect of the current-carrying conductor being in a magnetic field is that a force acts on it. If the force causes the conductor to move a distance δx in a time δt then there is an increase in flux linked by the circuit $\delta\Phi$ because of an increase in the circuit area of $L\delta x$ in the field. Thus

$$\delta\Phi = BL\delta x$$

Hence

$$\frac{\delta\Phi}{\delta t} = \frac{BL\delta x}{\delta t}$$

Thus if e is the induced e.m.f. as a result of this flux change,

$$e = -\frac{\delta\Phi}{\delta t} = -\frac{BL\delta x}{\delta t}$$

If the circuit has a resistance R, then when the source e.m.f. is E

$$E - \frac{BL\delta x}{\delta t} = IR$$

Multiplying both sides of the equation by I

$$EI - \frac{BIL\delta x}{\delta t} = I^2 R$$

$$EI\delta t = BIL\delta x + I^2 R\delta t$$

But $EI\delta t$ is the work done by the source in a time δt. The $I^2 R\delta t$ represents the heat generated in the circuit by the current. The remaining term, $BIL\delta x$, must represent the work done by the force used to move the conductor in the magnetic field. But the work done by a force F moving through a distance δx is $F\delta x$. Hence

$$F\delta x = BIL\delta x$$

$$F = BIL \qquad\qquad [42]$$

F is the force acting on a current-carrying conductor when in a magnetic field of flux density B at right-angles to its length and direction of motion. If the magnetic field is at some other angle to the current-carrying conductor then the component of the magnetic flux density at right-angles to the conductor is used.

A useful way of remembering the relationships between the various directions of current, field and force is *Fleming's left-hand rule* (Fig. 4.31). If the first finger represents the direction of the magnetic field, the second finger the current direction, and they are held at right-angles to each other and the thumb, then the direction of the thumb represents the direction of the force.

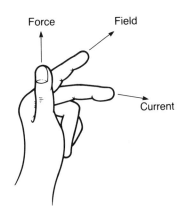

Force   Field

Current

Fig. 4.31 Fleming's left-hand rule

Example 28

What is the force acting per metre length of a conductor carrying a current of 5.0 A if it is at right-angles to a magnetic field having a flux density of 0.40 T?

Answer

Using the equation [42] developed above

$$F = BIL = 0.40 \times 5.0 \times 1 = 2.0 \text{ N/m}$$

Force on a current-carrying coil in a magnetic field

Figure 4.32 shows a single-turn, current-carrying, coil in a magnetic field. The vertical sides of the coil have a length L and the horizontal sides a length b. The horizontal sides are at an angle θ to the direction of the field, and the field has a uniform flux density B. The directions of the forces acting on the sides of the coil are: for the horizontal sides the forces are in opposite directions and since each side caries the same current these forces cause no motion of the coil; however the

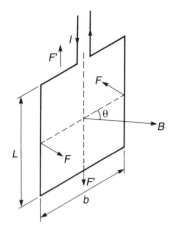

Fig. 4.32 Forces acting on a current-carrying coil in a magnetic field

forces acting on the vertical sides are also in opposite directions and the same size but they cause rotation. The force F acting on a vertical side is, by equation [42],

$$F = (B \cos \theta)IL$$

The turning moment or torque about the central vertical axis of the coil is thus

$$\text{torque} = (B \cos \theta)IL \times \tfrac{1}{2}b + (B \cos \theta)IL \times \tfrac{1}{2}b$$

$$= BILb \cos \theta$$

Lb is the area A of the loop. Hence

$$\text{torque} = BIA \cos \theta$$

If there are N turns on the coil then

$$\text{torque} = NBIA \cos \theta$$

This is the basis of the moving-coil meter. With such a meter the direction of the flux density is such that it is always at right-angles to the coil. Then

$$\text{torque} = NBIA \qquad\qquad [43]$$

This torque causes the coil to rotate against springs. The opposing torque developed by the springs is proportional to the angular deflection θ.

$$\text{Spring torque} = k\theta$$

where k is a constant for the springs. The coil thus rotates until the torque developed by the springs cancels that produced by the current through the coil. Then

$$k\theta = NBIA$$

$$\theta = NBIA$$

Hence the angular deflection is proportional to the current I. Because the angular deflection is proportional to the current the moving-coil meter has a linear scale.

Energy stored in the magnetic field of an inductor

When the current through an inductor changes an e.m.f. is induced which opposes the change producing it. If at some instant of time the current is changing at the rate di/dt, then the e.m.f. induced is by equation [14]

$$\text{e.m.f.} = -\frac{L \, di}{dt}$$

The value of the supply voltage v at this instant of time needed to overcome this and maintain the current through the inductor is thus

$$v = \frac{L\,di}{dt}$$

Energy is thus required to maintain the current and the power required is the product of the voltage that has to be applied and the current. Hence

$$\text{power } P = iv = iL\frac{di}{dt}$$

When the current is switched through an inductor it grows to reach a final steady value. At the steady value di/dt is zero and the power input becomes zero. The energy that has been supplied to the inductor has been used to establish the magnetic field around the inductor. In time dt the energy required is $P\,dt$, hence the energy required to increase the current from zero to I is

$$\text{energy} = \int_0^I Li\,di$$

$$= \tfrac{1}{2}LI^2 \qquad\qquad\qquad\qquad [44]$$

This is the energy stored in the magnetic field of the inductor. When the current through the inductor ceases the magnetic field collapses and the energy stored in the magnetic field is returned to the circuit and results in an e.m.f. being induced in it which continues to maintain the current for some time after the current supply to the inductor has been cut off.

Equation [44] can be expressed in other forms. Thus if the inductor is a solenoid, equation [16] gives

$$L = \frac{\mu N^2 A}{l}$$

where l is the length of the solenoid, A its cross-sectional area, N the number of turns and μ the permeability of the medium used for the core. Thus equation [44] becomes

$$\text{energy} = \frac{\mu N^2 A I^2}{2l}$$

The volume of the magnetic field is lA, hence

$$\text{energy/unit volume} = \frac{\mu N^2 I^2}{2l^2}$$

But the magnetic field strength H is NI/l (equation [5]) and since the flux density $B = \mu H$ (equation [6]), then B is $\mu NI/l$. Thus

$$\text{energy per unit volume} = \frac{B^2}{2\mu} \qquad\qquad\qquad [45]$$

or, since $B = \mu H$,

$$\text{energy per unit volume} = \tfrac{1}{2}BH \qquad [46]$$

Example 29

What is the energy stored in an inductor of inductance 200 mH when the current through it is 100 mA?

Answer

Using equation [44]

$$\text{energy} = \tfrac{1}{2}LI^2 = \tfrac{1}{2} \times 0.200 \times 0.100^2 = 1.0 \times 10^{-3} \text{ J}$$

Energy stored in the magnetic field of a magnetic circuit

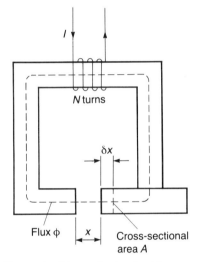

Fig. 4.33 Magnetic circuit with a changing air gap

An alternative way of considering the energy in a magnetic field is to consider a magnetic circuit. When a current flows through a coil it produces a magnetic field which results in forces acting on neighbouring pieces of iron or other current-carrying coils. Similarly a permanent magnet causes forces to act on neighbouring pieces of iron or current-carrying coils. When such pieces of iron or the current-carrying coils are made to move then energy is expended, work is done. This energy comes from the magnetic field. The above situations are essentially just a magnetic circuit in which the air gap in the circuit is changed, the air gap being the distance between the coil and the item which is moved. Figure 4.33 shows a representation of this situation.

The energy expended in producing the magnetic flux in the circuit when the current is i is the energy taken from the electrical source in raising the current in the coil from zero to the value i. Since the energy expended per second, when the current is i and the potential difference across the coil v, is vi then the energy expended in a time δt is $vi\,\delta t$. Hence the total energy taken from the electrical source, if a time t is taken to take the current from zero to i, is

$$\text{energy} = \int_0^t vi\ \mathrm{d}t$$

But the potential difference v across the coil is the value that is needed at some instant to maintain the current through the coil against the induced e.m.f., this being $-N\,\mathrm{d}\Phi/\mathrm{d}t$ (equation [2]). Hence v is $N\,\mathrm{d}\Phi/\mathrm{d}t$ and thus

$$\text{energy} = \int_0^t N\frac{\mathrm{d}\Phi}{\mathrm{d}t}\ i\ \mathrm{d}t$$

If at time t the flux has become Φ from being zero at time $t = 0$, then

$$\text{energy} = \int_0^t Ni\ \mathrm{d}\Phi$$

The quantity Ni is the m.m.f. (equation [4]). This equation can represent the energy in the entire magnetic field of the circuit if Ni is the m.m.f. of the coil, or the energy in a segment of the magnetic field if Ni is the m.m.f. required for that element of the field. Thus, for the air gap, since the magnetic field strength H in the air gap is Ni/x (equation [5]), then

$$\text{energy} = \int_0^\Phi Hx \, d\Phi$$

Since the flux density $B = \Phi/A$ (equation [3]), then $dB/d\Phi = 1/A$ and

$$\text{energy} = \int_0^B HxA \, dB$$

Since the volume of the magnetic field in the air gap can, in the absence of fringing, be assumed to be xA, then

$$\text{energy per unit volume} = \int_0^B H \, dB \qquad [47]$$

For a material such as air, where the permeability is a constant, we can substitute for H using $B = \mu H$ and obtain

$$\text{energy/unit volume} = \tfrac{1}{2}B^2/\mu \qquad [48]$$

or, since $B = \mu H$,

$$\text{energy/unit volume} = \tfrac{1}{2}BH \qquad [49]$$

or

$$\text{energy/unit volume} = \tfrac{1}{2}\mu H^2 \qquad [50]$$

The above equations give the energy stored in the magnetic field of the air gap.

To move the pole pieces of the magnetic circuit further apart and increase the air gap by δx forces have to be applied to pull them apart. The work done in increasing the separation by this distance against a force F is

$$\text{work done} = F\delta x$$

In increasing the separation the volume of the magnetic field in the air gap has been increased by $A\delta x$. Thus the extra energy stored in the magnetic field is, using equation [48],

$$\text{increase in magnetic field energy} = \tfrac{1}{2}B^2 A\delta x/\mu$$

But this increase must come from the input of mechanical energy involved in pulling the pole pieces apart. Thus

$$F\delta x = \tfrac{1}{2}B^2 A\delta x/\mu$$

Hence the force per unit surface area of a pole piece is

$$F/A = \tfrac{1}{2}B^2/\mu \qquad [51]$$

This equation gives the forces acting on the surfaces of the pole pieces in the magnetic circuit when there is an air gap between them.

Example 30

What is the energy stored in the iron and the air gap of a magnetic circuit of the form shown in Fig. 4.33 when the iron has a flux path of length 600 mm and the air gap a length 4.0 mm? The circuit has a cross-sectional area of 1000 mm². The m.m.f. of the circuit is provided by a coil of 200 turns through which a current of 2.0 A flows. The iron has a relative permeability of 2500.

$$\mu_0 = 4\pi \times 10^{-7} \text{ H/m}$$

Answer

Equation [10] gives for the reluctance S of an element

$$S = \frac{L}{\mu_r \mu_0 A}$$

Thus for the iron

$$S = \frac{0.600}{2500 \times 4\pi \times 10^{-7} \times A} = \frac{191}{A}$$

and for the air

$$S = \frac{0.004}{4\pi \times 10^{-7} \times A} = \frac{3183}{A}$$

Since the two reluctances are in series the total reluctance will be the sum of them, i.e., $3374/A$. The m.m.f. of the circuit is NI (equation [4]) and thus equation [9] can be used to obtain the flux in the circuit

$$\text{m.m.f.} = \Phi S$$

$$\Phi = \frac{200 \times 2}{3183/A}$$

The flux density is Φ/A (equation [3]) and thus

$$B = \frac{200 \times 2}{3183} = 0.13 \text{ T}$$

Thus for the iron, volume of field $0.600 \times 1000 \times 10^{-6}$ m³, the energy stored is, using equation [48],

$$\text{energy/unit volume} = \frac{1}{2} B^2 / \mu$$

$$\text{energy} = \frac{0.13^2 \times 0.600 \times 1000 \times 10^{-6}}{2 \times 2500 \times 4\pi \times 10^{-7}} = 1.6 \times 10^{-3} \text{ J}$$

For the air, volume $0.004 \times 1000 \times 10^{-6}$ m², the energy stored is

$$\text{energy} = \frac{0.13^2 \times 0.004 \times 1000 \times 10^{-6}}{2 \times 4\pi \times 10^{-7}} = 2.7 \times 10^{-2} \text{ J}$$

Example 31

What are the forces acting on the surfaces of two pole pieces in a magnetic circuit if the air gap between them has a constant cross-sectional area of 500 mm^2 and the flux in the air gap is 100 μWb?

$$\mu_0 = 4\pi \times 10^{-7} \text{ H/m}$$

Answer

Using equation [50]

$$F/A = \tfrac{1}{2}B^2/\mu$$

and, since $B = \Phi/A$ and the medium is air,

$$F = \frac{\Phi^2}{A\mu_0}$$

$$= \frac{(100 \times 10^{-6})^2}{500 \times 10^{-6} \times 4\pi \times 10^{-7}} = 16 \text{ N}$$

Problems

1 What is the e.m.f. induced in an air-cored coil of 400 turns if the flux linked by it changes from 20 μWb to 50 μWb in 1.4 ms?

2 What is the m.m.f. produced by a coil of 800 turns when it is carrying a current of 3 A?

3 What is the magnetic field strength at the centre of a coil of 1000 turns and length 200 mm when carrying a current of 5.0 A?

4 What is the (*a*) reluctance of a magnetic circuit, (*b*) the flux in the magnetic core, if the circuit is wound with a coil of 1000 turns carrying a current of 500 mA, has a mean flux path length of 300 mm, a cross-sectional core area of 1000 mm^2 and a core with a relative permeability of 100?

5 A coil of 400 turns is uniformly wound on a wooden core, relative permeability 1, having a mean circumference of 700 mm and a uniform cross-sectional area of 450 mm^2. What is (*a*) the magnetic field strength, (*b*) the flux density and (*c*) the flux in the core when a current of 3 A passes through the coil?

6 What is the m.m.f. needed to produce a flux density of 0.50 T in an air gap 2.0 mm long?

7 A coil of 300 turns is uniformly wound on an iron ring with a mean circumference of 450 mm and a cross-sectional area of 400 mm^2. What is (*a*) the reluctance of the ring and (*b*) the current required to produce a flux density of 1.5 T in the ring? The relative permeability of iron at this flux density may be taken as 350.

8 An iron ring uniformly wound with a coil of 500 turns and having a mean circumference of 400 mm and a cross-sectional area of 350 mm^2 is cut radially so that an air gap of length 2 mm is introduced. What is (*a*) the reluctance of the iron core, (*b*) the reluctance of the air gap, (*c*) the current required to produce a flux density of 2.0 T in the air gap? The relative permeability of the iron may be taken as 400 and it may be assumed that no magnetic leakage of fringing occurs.

9 An electromagnet has an iron core with a mean flux path of length 800 mm and is wound with a coil of 800 turns. What must be the current in the coil if the electromagnet is to produce a flux density of 0.70 T in its 2 mm air-gap? The relative permeability of the iron at this flux density may be taken as 400 and any effects of magnetic leakage or fringing neglected.

10 A magnetic circuit, of the form shown in Fig. 4.11, has a mean flux path in the iron of 300 mm and an air gap of length 4 mm. The cross-sectional area of the iron is 1000 mm². What is the flux in the air gap when the circuit is supplied with an m.m.f. of 3000 A? Take the relative permeability of the iron to be 100.

11 A magnetic circuit of the form shown in Fig. 4.12 has a central limb which has twice the cross-sectional area of the two outer limbs, they having equal cross-sectional areas of 400 mm². The central limb is wound with 200 turns of wire carrying a current of 2.0 A. The length of the flux path through the central limb and one outer limb is 250 mm. How does (*a*) the flux and (*b*) the flux density in the central limb compare with that in an outer limb and (*c*) what is the flux density in the central limb? The following is *B–H* data for the circuit core.

H (A/m)	1200	1400	1600	1800
B (T)	0.31	0.37	0.42	0.47

12 An iron rod of length 600 mm is bent into a ring with a 2 mm air gap between the ends of the rod. The cross-sectional area of the rod is 400 mm². The rod is uniformly wound with a coil of 900 turns. What should be the current in the coil if the magnetic flux in the air gap is to be 0.2 mWb? The following is *B–H* data for the iron. Neglect any magnetic leakage.

H (A/m)	90	100	110	120	140
B (T)	0.43	0.47	0.50	0.53	0.60

13 What would the current be for problem 12 if the magnetic leakage coefficient is 1.2?

14 What is the average e.m.f. induced in a coil of inductance 10 mH when the current changes from 10 mA to 40 mA in 5 μs?

15 What is the inductance of an air-cored coil of 1000 turns if a current of 5 A through it produces a magnetic flux of 180 μWb?

16 What is the inductance of an air-cored coil of 600 turns if it has a length of 200 mm and a cross-sectional area of 300 mm²?

17 What is the inductance per metre length of a coaxial cable which has an inner core of radius 2.0 mm and an outer conductor of internal radius 8.0 mm. Assume that the relative permeability of the medium between the conductors is 1. Neglect any inductance due to internal linkages within the inner core.

18 What will be the inductance per metre length for the cable specified in problem 17 if internal linkages are not neglected?

19 What will be the inductance per metre length of a pair of parallel cables in air, each cable with radius 5.0 mm, when they are 500 mm apart? Neglect any inductance due to internal linkages.

20 If two coils have a mutual inductance of 200 μ*H*, what will be the e.m.f. induced in one coil when the current in the other coil changes at the rate of 20 000 A/s?

21 What is the coupling coefficient for a pair of coils if they have inductances of 12 mH and 24 mH and a mutual inductance of 6 mH?

22 A single-phase transformer has a step-down ratio of 10:1. (*a*) How many turns will there be on the secondary coil if the primary has 200 turns? (*b*) What will be the primary current if the secondary current is 30 A?

23 A single-phase transformer has 600 primary turns and 1000 secondary turns. What will be (*a*) the maximum value of the flux produced in the transformer core, (*b*) the secondary induced e.m.f. if the input to the primary turns is 240 V, 50 Hz?

24 Sketch the phasor diagram for a transformer on no load and explain the term no-load current and its components the magnetising current and core-loss current.

25 Explain what is meant by a transformer having leakage flux and how such flux can be minimised in the design of transformers.

26 Sketch the equivalent circuit of a single-phase transformer that includes the effects of resistance and leakage reactance.

27 A transformer has 600 primary turns and 120 secondary turns. The primary resistance is 0.42 Ω and the secondary 0.02 Ω. The primary leakage reactance is 1.30 Ω and the secondary 0.04 Ω. What is the equivalent impedance which could be placed in the primary circuit with an ideal transformer?

28 What should be the turns ratio of a transformer which is to be used to match an amplifier with an output impedance of 100 Ω with a circuit of input impedance 50 Ω and so give maximum power transfer?

29 What is the efficiency of a 10 kVA, single-phase transformer, operating at half load if it has an iron loss of 100 W and a full-load copper loss of 160 W and the power factor of the load is 0.6?

30 What is meant by the terms iron and copper losses?

31 A transformer has an eddy current loss of 12 W when operating at a frequency of 100 Hz. What will be the eddy current loss when it is used at 50 Hz, the flux density remaining unchanged?

32 When the maximum flux density in an inductor is 1.2 T the hysteresis loss is 4.0 W. What will be the loss when the maximum flux density is 1.6 T? The Steinmetz constant is 1.6.

33 The iron losses of a transformer are measured at different frequencies and a graph plotted of (iron loss/frequency) against the frequency. If the graph has an intercept with the (iron loss/frequency) axis of 0.20 W/Hz and a slope of 0.010 W/Hz^2, what will be the hysteresis and eddy current losses at 100 Hz?

34 What is the force acting per metre length of a conductor carrying a current of 2.0 A if it is (*a*) at right angles, (*b*) parallel to a magnetic field having a flux density of 0.60 T?

35 What is the energy stored in the magnetic field of an inductor of inductance 300 mH when a current of 100 mA flows through it?

36 What are the forces that need to be applied to maintain an air gap between two pole pieces in a magnetic circuit if the flux in the air gap is 120 μWb and the cross-sectional area of each pole piece is 600 mm^2? $\mu_0 = 4\pi \times 10-7$ H/m.

5 Magnetic materials

Why are some materials very magnetic and others not? What materials are suitable for permanent magnets? What materials are suitable for the cores of transformers? This chapter is concerned with considering the atomic basis of magnetism, the criteria which are used to specify the magnetic properties of materials and then the significance of these criteria in relation to the selection of materials for permanent magnets and transformer cores.

Dipole moment

The maximum torque acting on a coil in a magnetic field is given by (see Ch. 5 eqn [43])

$$\text{maximum torque} = NIAB$$

where N is the number of turns, A the cross-sectional area of the coil, I the current through the coil and B the flux density. The maximum value is when the magnetic flux density is at right-angles to the coil. See Chapter 4 for the derivation of this equation. The quantity NIA is called the *magnetic dipole moment* of the coil. Thus

$$\text{magnetic dipole moment} = NIA \qquad [1]$$

and

$$\text{maximum torque} = \text{magnetic dipole moment} \times B \qquad [2]$$

The bigger the dipole moment of a coil the greater the maximum torque for a given magnetic field. The units of magnetic dipole moments are A m^2.

The electrons in an atom can be considered to be in orbit round the nucleus (see Ch. 1). For one electron in such an orbit this is rather like a current in a single-turn coil and so we can consider there to be a magnetic dipole moment of IA. In addition to orbiting a nucleus the electron can also be considered to be spinning. If we consider the electron to be a small charged sphere then the rotation of the charge on the surface of the sphere at the equators will be like a single-turn current loop and also produce a magnetic moment. Figure 5.1 indicates how we can think of these magnetic dipole moments being produced and the directions of the flux produced by them.

(a)

(b)

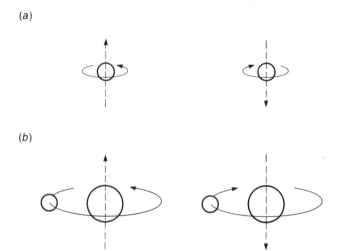

Fig. 5.1 Magnetic dipole moments produced by (a) electron spin, (b) electron orbits around the nucleus

In a block of material there will be many atoms and so it is useful to talk of the magnetic dipole moment per unit volume M.

$$M = \frac{\text{sum of all atomic dipole moments}}{\text{volume of the material}} \qquad [3]$$

The units of M are A m^2/m^3 = A/m.

Intensity of magnetisation

The flux density B in a material can be regarded as the sum of the flux density that would have occurred in a vacuum (free space) of $\mu_0 H$, where H is the applied magnetic field strength, and the flux density due to the magnetisation of the material J.

$$B = \mu_0 H + J \qquad [4]$$

J is called the *magnetic polarisation* and has the same unit as flux density, i.e., tesla. This flux density is due to the magnetic dipoles in the material. The magnetic dipole moment per unit volume M can be considered to be the internally produced magnetic field strength, and is thus often called the *intensity of magnetisation*. Thus

$$J = \mu_0 M \qquad [5]$$

and so

$$B = \mu_0 H + \mu_0 M = \mu_0(H + M)$$

The relative permeability μ_r is given by $B = \mu_0\mu_r H$, hence

$$\mu_r = \frac{H + M}{H} = 1 + \frac{M}{H} = 1 + \chi \qquad [6]$$

The term M/H is called the *magnetic susceptibility* χ. It states how susceptible to magnetisation a material is, the higher its value the more susceptible a material is and the higher the relative permeability. It is a ratio and has no units.

Example 1

A magnetic field of 2000 A/m is applied to a material which has a susceptibility of 1200. What will be (*a*) the relative permeability, (*b*) the intensity of magnetisation and (*c*) the flux density?

Answer

(*a*) Using the equation [6] given above

$$\mu_r = 1 + \chi = 1 + 1200 = 1201$$

(*b*) Since $\chi = M/H$, then

$$M = \chi H = 1200 \times 2000 = 2.4 \times 10^6 \text{ A/m}$$

(*c*) Since $B = \mu_r\mu_0 H$, then

$$B = 1201 \times 4\pi \times 10^{-7} \times 2000 = 3.0 \text{ T}$$

Atomic magnetic dipole moments

Each electron in an atom can be considered to have two ways of contributing to the magnetic dipole moment, one is by its orbital motion round the nucleus and the other is by its spin. The magnetic quantum number m_l (see Ch. 1) is involved in determining what values the magnetic dipole moment can have due to electron rotation about the nucleus.

$$\text{Orbital magnetic moment} = -m_l\beta \qquad [7]$$

where β is called the *Bohr magneton* and has the value of 9.27×10^{-24} A m^2. For an atom m_l has integer values between $-l$ and $+l$, with l being the angular momentum quantum number.

With regard to spin the spin quantum number m_s is involved. For each electron m_s is either $+\frac{1}{2}$ or $-\frac{1}{2}$. These two possibilities represent the two different directions of spin illustrated in Fig. 5.1(*a*).

$$\text{Spin magnetic moment} = -m_s(2\beta) \qquad [8]$$

Thus an electron in an atom can have a spin magnetic moment of $+1$ Bohr magneton or -1 Bohr magneton.

The total magnetic dipole moment of an atom depends on the sum of the contributions made by each electron and the extent to which they cancel out because they are in opposite directions. For atoms in solids the main source of magnetism is the spin motion of the electrons since the dipole moments produced by orbital motions tend to cancel each other out. The spin motions of electrons in filled orbitals cancel each other out since there are as many electrons spinning in one direction as in the opposite direction. In a solid valence electrons from one atom usually bond with valence electrons from other atoms so that an electron spinning in one direction is paired with one spinning in the opposite direction and so their magnetic moments cancel out. Thus for most solids the atomic magnetic dipole moments cancel each other out.

Types of magnetic materials

Materials can be grouped into three general categories:

1 *diamagnetic materials* which have relative permeabilities slightly below 1;
2 *paramagnetic materials* which have relative permeabilities slightly greater than 1;
3 *ferromagnetic and ferrimagnetic materials* which have relative permeabilities considerably greater than 1. Ferromagnetic materials are metals while ferrimagnetic materials are ceramics.

The relative permeability of a vacuum is 1, thus a graph of flux density B against magnetising field strength H for paramagnetic, diamagnetic materials and a vacuum looks like Fig. 5.2(*a*). Since the relative permeability is reasonably independent of the value of B the graphs are straight lines with a constant slope. Ferromagnetic and ferrimagnetic materials have large relative permeabilities which depend on the value of B, thus Fig. 5.2(*b*) shows the line indicating the maximum value of the relative permeability.

Fig. 5.2 *B–H* graphs for (*a*) paramagnetics, diamagnetics and a vacuum, (*b*) ferromagnetics

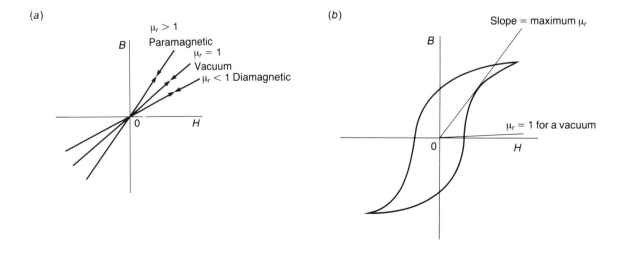

(*a*)

$\mu_r > 1$
B Paramagnetic
$\mu_r = 1$
Vacuum
$\mu_r < 1$ Diamagnetic
0   H

(*b*)

Slope = maximum μ_r
B
$\mu_r = 1$ for a vacuum
0   H

Diamagnetic materials

Diamagnetic materials have a relative permeability less than 1 and the susceptibility is negative. They have atoms which do not have permanent dipole moments. We can consider the orbit of each electron about a nucleus to give rise to a dipole moment. However, the magnetic effects tend to cancel each other out and there is no net magnetic moment for atoms in a solid. When a magnetic field is applied electromagnetic induction (see Ch. 4) can occur for each of the orbiting electrons since each can be considered as effectively a single-turn coil carrying a current. The induced e.m.f. results in an induced current which is in such a direction as to oppose the change producing it (see Ch. 4). This means that the magnetic field produced by the induced current in the atomic coils is in the opposite direction to the magnetic field responsible for producing it. The result of applying the magnetic field to a diamagnetic material is thus to produce temporary dipole moments, these vanishing when the magnetic field is removed. Because these dipole moments oppose the field producing them they reduce the flux density to a value less than it would otherwise have had in a vacuum. Bismuth, copper, mercury and water are examples of diamagnetic materials.

All materials have diamagnetism since all can be considered to have orbiting electrons. But for diamagnetic materials there is no contribution to the magnetism from the spin of the electrons, while for paramagnetic and ferromagnetic there is. The magnetic dipole moments produced as a result of spin in such materials completely swamps the small diamagnetic effect.

Paramagnetic materials

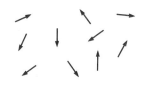

Fig. 5.3 Dipole arrangements in paramagnetic materials

Paramagnetic materials have a relative permeability only slightly greater than 1, and the susceptibility is positive and small. These have atoms or ions with a net dipole moment. The magnetic dipoles are, in the absence of an applied magnetic field, completely randomly orientated (Fig. 5.3) and thus the material shows no permanent magnetism. When a magnetic field is applied some small degree of alignment in the field direction can occur. The effect vanishes when the field is removed. The materials are thus only weakly magnetic in the presence of a field. Aluminium, platinum and solid oxygen are examples of paramagnetic materials.

Ferromagnetic and ferrimagnetic materials

Ferromagnetic materials have very high values of relative permeability and high positive values of susceptibility. The atoms in these solids have dipole moments, with the way in which the atoms bond together to form the solid resulting in the dipoles in neighbouring atoms aligning themselves all in

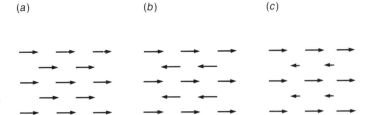

Fig. 5.4 Dipole arrangements in (a) ferromagnetics, (b) antiferromagnetics, (c) ferrimagnetics

the same direction (Fig. 5.4(a)). It is this coupling between atoms which results in their magnetic dipoles aligning that distinguishes ferromagnetic materials from paramagnetic materials where the dipoles are randomly orientated. Examples of ferromagnetics are iron, cobalt and nickel.

Some materials have dipoles which though they align themselves have half pointing in one direction and half in the exactly opposite direction (Fig. 5.4(b)). Such a material is called *antiferromagnetic* when the dipoles are the same size and cancel each other out. In some cases the dipoles are not all the same size and do not cancel each other out (Fig. 5.4(c)) or a solid is a mixture of different atoms for which some have dipoles which cancel out and others do not, such materials being called *ferrimagnetics* or *ferrites*. Examples of antiferromagnetics are manganese, manganese oxide and chromium; and of ferrites iron oxide Fe_3O_4, nickel ferrite $NiFe_2O_4$ and manganese ferrite $MnFe_2O_4$. Ferrites are essentially ceramics and so are brittle and difficult to machine.

Atomic structure of ferromagnetics

Electrons in filled orbits of atoms consist of equal numbers of electrons spinning in one direction and in the opposite direction, thus their dipole moments cancel each other out. Some of the electrons in the outer unfilled orbit of an atom can be unpaired; however these are the electrons involved in the bonds between atoms and in the case of a metallic bond (see Ch. 1) can become detached or in the case of other bonds end up becoming paired with an electron in a neighbouring atom. Bonding thus tends to mop up the unpaired electrons. However, certain elements have inner orbitals which are not completely filled. The transition elements, scandium to zinc, are such elements (see Ch. 1 for the arrangement of electrons in atoms of these elements). Table 5.1 shows details of the directions of the spin of their 3d electrons, this being the partially filled orbital. Atoms of most of these elements have some unpaired spins. But not all of them in the solid form are ferromagnetic. This is because, for some, when the atoms bond together for the solid the magnetic moment of one atom

Table 5.1 Electron spin directions for transition elements

Element	Number of 3d electrons	Spin directions of electrons
Scandium	1	1 upwards
Titanium	2	2 upwards
Vanadium	3	3 upwards
Chromium	5	5 upwards
Manganese	5	5 upwards
Iron	5	5 upwards 1 downwards
Cobalt	7	5 upwards 2 downwards
Nickel	8	5 upwards 3 downwards
Copper	10	5 upwards 5 downwards
Zinc	10	5 upwards 5 downwards

may cancel out that of a neighbouring atom or the bonding between the atoms may pair up the spins from neighbouring atoms or the atom may lose the unpaired electron.

The atoms of iron, cobalt and nickel in the sold state are able to retain some unpaired spins, iron for example retaining on average 2.2 unpaired electrons per atom with the others breaking 'free' to form the metallic bond (see Ch. 1). The bonding between the atoms in these solids is such that the magnetic moments of neighbouring unpaired electrons tend to align in the same direction. The result is a net magnetic dipole moment for the solid. Such materials are said to be *ferromagnetic*.

A ferromagnetic solid has thus:

1 atoms with unpaired spins in the solid (isolated atoms may have unpaired spins but the bonding of the atoms together to form the solid can result in spins of neighbouring atoms becoming paired or the electrons becoming detached from the atom as part of the bonding process);
2 bonding between atoms which results in neighbouring atoms having magnetic moments in the same direction.

Example 2

Though manganese has five unpaired electrons per atom it is not in the solid strongly magnetic, why not?

Answer

Though each manganese atom has unpaired electrons when the atoms bond together in a solid the magnetic moments in neighbouring atoms cancel each other out and so the material is not strongly magnetic, i.e., ferromagnetic. Manganese is antiferromagnetic.

Atomic structure of ferrimagnetic materials

Ferromagnetic materials are metals with the bonds between atoms being metallic bonds, i.e., bonds resulting from free electrons (see Ch. 1). Ferrimagnetic materials are ceramics with ionic bonds (see Ch. 1), i.e., bonds resulting from the electric forces of attraction between positive and negative ions. A major consequence of this difference in bonding between ferromagnetics and ferrimagnetics is that ferromagnetics are good conductors of electricity with very low resistivities while ferrimagnetics are poor conductors with much higher resistivities.

The naturally occurring mineral magnetite Fe_3O_4, known as lodestone, is a ceramic magnetic material. It is a crystalline structure composed of Fe^{2+}, Fe^{3+9} and O^{2-} ions. The Fe^{2+} ions are iron atoms that have lost two electrons, those lost being two 4s electrons. This does not affect the four unpaired 3d electrons and so the ion has a dipole moment. The Fe^{3+} ions are atoms that have lost three electrons, those lost being the two 4s electrons and one of the 3d electrons, leaving five unpaired electrons and hence a magnetic moment. The oxygen ions have no dipole moment since they have no unpaired electrons. The way the ions are arranged in the crystal leads to half the Fe^{3+} ions lining up in one direction and half in the other with the result that there is no net dipole moment due to them, such behaviour being referred to as antiferromagnetic. The Fe^{2+} ions however interact with each other to line up in the same direction and so give a net magnetic moment for the material. This type of structure is typical of ferrites.

Modern ferrite soft magnetic materials (see later in this chapter for an explanation of the term soft) are produced by having the same crystalline structure as the magnetite but replacing the Fe^{2+} ions with other ions, e.g., Mn^{2+} ions, and/or replacing some of the Fe^{3+} ions with other ions, e.g., Zn^{2+}. The Mn^{2+} ions have five unpaired 3d electrons and line up in the same direction in the same way as the Fe^{2+} ions. The Zn^{2+} ions, however, have no unpaired electrons and no dipole moment. However, they replace some of the unproductive Fe^{3+} ions and force some of them to contribute to the net magnetic moment.

Permanent magnetic ferrite materials, referred to as hard magnetic materials (see later in this chapter for an explanation of the term), have a slightly different form to their crystalline structure but follow the same basic principles. Thus one consists of positive barium and iron ions with negative oxygen ions, while another has positive lead and iron ions with negative oxygen ions.

Curie temperature

With a ferromagnetic material the spins of electrons in neighbouring atoms tend to be coupled so that they spin in the same direction and their magnetic dipole moments add up rather than cancel. However, this coupling can be upset by a high enough temperature. This is because thermal energy shakes the coupled spins out of alignment. The temperature at which all alignment is lost is called the *Curie temperature*. For iron the Curie temperature is 770 °C, for nickel 358 °C and cobalt 1115 °C. Thus to be of use as a ferromagnetic material the Curie temperature has to be taken into account and the materials only used below that temperature.

Example 3

A magnetic material, a silicon steel, is specified as having a Curie temperature of 745 °C. How will the properties of the material below this temperature differ from above it?

Answer

Below the Curie temperature the material will be highly magnetic, a ferromagnetic material, above it only very weakly magnetic.

Domains

The dipoles in ferromagnetic and ferrimagnetic materials are not generally all arranged in lines in the same direction but in blocks or *domains*. Within a domain all the dipoles have the same alignment but the direction of the alignment of different domains may differ and not be in the same direction. Figure 5.5(*a*) shows the type of structure a ferromagnetic material may have in the absence of an applied field. The directions of the domains may be completely random and so the material shows no permanent magnetism. When a field is applied, those domains most nearly in the direction of the field grow in size at the expense of neighbouring wrongly directed domains (Fig. 5.5(*b*)) and rotate to line up with the field. Increasing the applied field increases the growth of the appropriately aligned domains. The result is a greater magnetisation of the material. When all the domains are in the direction of the applied field the material is said to be *saturated*. The material is then

(*a*)   (*b*)

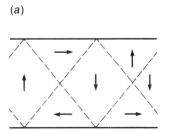 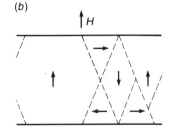

Fig. 5.5 Domains when there is (*a*) no applied field, and when (*b*) a field is applied and domain growth occurs

showing its greatest possible amount of magnetisation. When the applied field is removed many of the domains remain orientated in the same direction and thus the material retains a residual magnetism.

The hysteresis loop

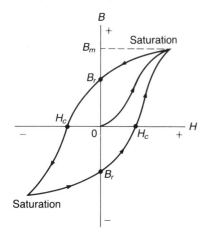

Fig. 5.6 The hysteresis loop for a ferromagnetic material

Figure 5.6 shows how the flux density B in a ferromagnetic material changes as the magnetic field strength H is changed. Thus starting at point 0 when all the domains are randomly orientated, increasing H results in an increase in B as domains in the direction of the field grow. This continues until all the domains are in the same direction and the material is *saturated*. The flux density is then a maximum, B_m. When the field strength H is reduced to zero the domains do not all become unaligned and the material retains a permanent flux density B_r, known as the *remanence*. Reversing the direction of the applied field starts the growth of domains in this reverse direction. At field strength H_c, called the *coercive field*, the flux density is reduced to zero since then there are as many domains in one direction as another. Further increase in this reverse direction field strength results in further growth of the domains in the field direction until saturation in the reverse direction is achieved. When this field is then reduced to zero a residual flux density B_r in this reverse direction remains. Applying the field in the initial direction results in domains aligned in that direction growing until saturation is again achieved and the cycle is complete.

Hysteresis loss

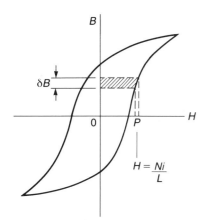

Fig. 5.7 Hysteresis loss

Consider the hysteresis loop (Fig. 5.7) obtained for an iron ring of mean circumference L and cross-sectional area A and wound with a coil of N turns. If the current through the coil at some instant is i then the magnetic field strength H is (eqn [5] Ch. 4)

$$H = \frac{Ni}{L}$$

Suppose this magnetic field strength to correspond to point P on the B–H graph. If now the magnetic field strength is increased by a small amount in a time δt the flux density increases by a small amount δB. The e.m.f. induced in the coil from this change is, using equation [2] Chapter 4,

$$\text{e.m.f.} = -\frac{N\,\mathrm{d}\Phi}{\mathrm{d}t} = -\frac{N\,\mathrm{d}(BA)}{\mathrm{d}t} = -\frac{NA\delta B}{\delta t}$$

The applied voltage v needed to neutralise this e.m.f. is

$$v = \frac{NA\delta B}{\delta t}$$

Therefore the power supplied at this instant to the magnetic field is

$$\text{power } p = iv = \frac{iNA\delta B}{\delta t}$$

The energy that has to be supplied in the time δt is thus

$$\text{energy} = p\delta t = iNA\delta B$$

But $H = Ni/L$, hence

$$\text{energy} = H\delta B \times AL$$

AL is the volume of the material, hence the energy supplied per unit volume is

$$\text{energy/volume} = H\delta B$$

But $H\delta B$ is the area of the strip shown shaded in the hysteresis loop in Fig. 5.7. We can consider the entire loop divided into small strips and so the total energy needed to complete a loop is

$$\text{energy/volume} = \text{area enclosed by the hysteresis loop} \quad [9]$$

This is the energy loss during one cycle of the magnetising current. Steinmetz found that the area enclosed by the hysteresis loop is proportional to the maximum flux density to some power n, where n has a value depending on the quality of the ferromagnetic material and the range of flux density over which it is used. See Chapter 4 for further discussion.

Chapter 4 includes a more detailed consideration of the energy stored in the magnetic field.

Example 4

Estimate the hysteresis loss per cycle for the material giving the hysteresis loop shown in Fig. 5.8.

Answer

The loop is roughly a rectangle with a height of 0.32 T and width 140 A/m. The area of the loop is thus about 44.8 T A/m or 44.8 J. This is the energy loss per unit volume of the material during one cycle.

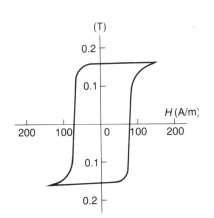

Fig. 5.8 Example 4

Soft and hard magnetic materials

The term *soft* when applied to magnetic materials comes originally from the magnetic properties associated with pure iron. Such a material was mechanically soft. However, now the term is applied to those magnetic **materials which,**

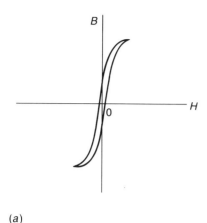

(a)

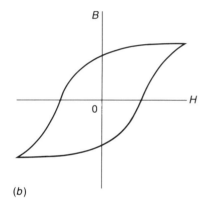

(b)

Fig. 5.9 Hysteresis loops (a) soft magnetic materials, (b) hard magnetic materials

regardless of whether they are mechanically soft or not, have:

1 high permeability, so that high magnetisation can be achieved;
2 a low coercive field, so that only a small magnetic field strength is required to demagnetise or reverse the direction of the magnetic flux in the material;
3 a small remanence, so that only a little magnetism is retained in the absence of a field;
4 a small area enclosed by the hysteresis loop so that little energy is lost per cycle.

The hysteresis loop in Fig. 5.9(a) is for a soft magnetic material. Such materials are widely used for electrical applications, e.g., transformer cores. Table 5.2 gives typical data for some commonly used soft magnetic materials. A typical transformer material would be the Fe–3% silicon alloy (see later this chapter for further discussion of this use).

The magnetic materials required for permanent magnets require different properties. They have an hysteresis curve looking like that shown in Fig. 5.9(b). Such a material is known as a *hard* magnetic material, because originally it applied to alloys of iron which were mechanically hard. Now, regardless of whether a material is mechanically hard or not a material is said to be magnetically hard if it has the properties of:

1 high remanence so that a high degree of magnetism is retained in the absence of a magnetic field;
2 high coercive field so that it is difficult to demagnetise the material;
3 a large area enclosed by the hysteresis loop, this being a consequence of it having a high remanence and high coercive force and this is a vital feature since high energy is then needed to demagnetise.

Table 5.2 Properties of soft magnetic materials

Material	Initial μ_r	Max. μ_r	B_r T	H_c A/m	ϱ $\Omega\,m$
Fe–3% Si (silicon steel)		90 000		6	5×10^{-7}
		8 000		40	5×10^{-7}
Fe–70/80% Ni (e.g., Mumetal)	60 000		0.5	1	6×10^{-7}
Fe–24% Co (Permendur)	300		1.7	950	2×10^{-7}
Ni–Zn ferrite	20–600				1000–300
Mn–Zn ferrite	600–5000				5–0.5

Note B_r is the remanence, H_c the coercive field and ϱ the electrical resistivity. The first entry for silicon steel is where the material has been rolled to give a strongly orientated crystalline structure with the specified properties in the direction of the rolling, the second entry is for non-orientated silicon steel.

See Table 5.3 later in this chapter for data for hard magnetic materials.

There is another shape of hysteresis loop which is of particular significance, a hysteresis loop which is virtually a square or rectangle (as in Fig. 5.8). This means that when the magnetising field is removed there is virtually no drop in the flux density, the saturation and the remanence flux densities being virtually the same. This type of loop is particularly useful for materials used for 'memories', e.g., the material used for computer floppy disks or that for the magnetic tape for tape recorders. The material is magnetised and then retains most of the flux density when the magnetising field is switched off.

The differences in properties between soft and hard magnetic materials can be explained in terms of the behaviour of the domains. Soft magnetic materials have domains which can easily grow and rotate. Thus when an applied magnetic field is removed the domains easily become disorientated and so the material has little permanent magnetism. Hard magnetic materials have domains which are locked into position and so when the applied magnetic field is removed the domains tend to stay in their fixed positions and so the material retains its magnetism. Any feature which produces disorder in the regular arrangement of atoms in the material will reduce the freedom of domains to move. Such features may be particles of a second material, dislocations in the orderly array of atoms, the boundaries between the different crystals which form the solid. Thus a hard magnetic material might be a mixture of small crystals and particles of a variety of different materials while a soft material is likely to be a less complex mixture and composed of larger crystals.

Example 5

For the following applications, which require soft magnetic materials and which hard?
(a) The core of an electromagnet.
(b) A compass needle.
(c) A magnetic door latch.

Answer

(a) A soft material is required since when the current through the coil surrounding the core is switched off the electromagnet should lose its magnetism.
(b) A hard material since the needle should be a permanent magnet which retains its magnetism.
(c) A hard material since the door latch should be a permanent magnet and retain its magnetism.

Permanent-magnet design

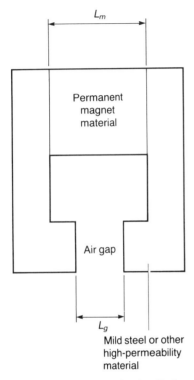

Fig. 5.10 The magnetic circuit of a permanent magnet

Consider the permanent magnet shown in Fig. 5.10. It is a magnetic circuit consisting of a block of permanent-magnet material, i.e., hard magnetic material, in series with soft magnetic material and an air gap. The flux through the permanent-magnet material is $B_m A_m$, where B_m is the flux density in the magnetic material and A_m its cross-sectional area. This, in the absence of leakage, would be the flux throughout the circuit and hence through the air gap. If however there is a leakage factor k_1, then

$$B_m A_m = k_1 B_g A_g$$

where k_1 is the leakage factor, B_g the flux density in the air gap and A_g the cross-sectional area of the air gap. The value of k_1 depends on the form of the magnetic circuit and typically has values between 1 and 20. For example, the magnetic circuit in a loudspeaker is likely to have k_1 values between 2 and 3.

The magnetomotive force for a length of flux path L is HL, where H is the magnetic field strength for that path. For the flux path through the magnet the m.m.f. is thus $H_m L_m$, where L_m is the length of the flux path in the magnet and H_m the magnetic field strength. The magnetic circuit does not have any current-carrying coil wrapped round it, thus we must have

m.m.f. for air gap + m.m.f. for magnet = 0

Thus the m.m.f. for the magnet would be the m.m.f. across the air gap if the connecting pieces and the joints between them and the permanent-magnet material had infinite permeability and hence zero reluctance. We can allow for this not being perfectly the case by introducing a correction factor k_2 in the equation. The value of k_2 typically lies between 1.1 and 1.3.

$$H_m L_m = k_2 H_g L_g$$

L_g is the length of the air gap and H_g the field strength in the gap. Since $B_g = \mu_0 H_g$, then

$$H_m L_m = k_2 (B_g / \mu_0) L_g$$

Multiplying this equation by our earlier equation,

$$B_m H_m A_m L_m = k_1 k_2 B_g^2 L_g A_g$$

The volume of the magnetic material V_m is $A_m L_m$ and the volume of the air gap V_g is $A_g L_g$, hence

$$V_m = \frac{k_1 k_2 B_g^2 V_g}{B_m H_m \mu_0} \qquad [10]$$

The volume of the magnetic material to achieve the required flux density B_g will be a minimum when $B_m H_m$ is a maximum.

Thus in permanent-magnet design an important parameter is the maximum value of the product *BH*.

As indicated by equation [49] in Chapter 4, the energy stored per unit volume of a magnetic field is ½*BH*, thus the maximum value of *BH* is a measure of the maximum amount of magnetic energy that can be stored in the magnetic field. Hence equation [10] indicates the conditions necessary to obtain the maximum amount of energy.

Figure 5.11 shows the negative magnetic field strength quadrant of the hysteresis loop for a permanent-magnet material. It is this quadrant, the demagnetisation curve, which is generally of interest with permanent magnets. This tells us how well a permanent magnet is able to retain its magnetism. Corresponding to this a *B–BH* graph has been drawn. The maximum value of (*BH*) occurs at *B* and *H* values corresponding to the largest rectangle which can be drawn in the quadrant. For a large value of the product *BH* there needs to be a large value of the coercive field and the remanence.

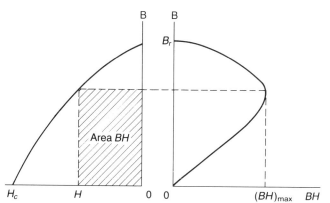

Fig. 5.11 The *BH–B* graph and its related *B–H* graph

Table 5.3 shows the remanence B_r, the coercive field H_c and the maximum value of the *BH* product for a number of hard magnetic materials. The main materials used for permanent magnets are the iron–cobalt–nickel–aluminium alloys, ferrites, and rare earth alloys. Steels, which used to be widely used, are not so often used since they have inferior properties to the other elements. Many magnetic materials are made *anisotropic*, i.e., they have magnetic properties in one direction considerably better than in other directions. The Fe–Co–Ni–Al, ferrites and rare earth alloys are generally anisotropic.

The Fe–Co–Ni–Al alloys have high remanence, a high maximum *BH* value and moderately high coercivity. They have a high Curie temperature and can be used at temperatures of 500 °C and higher. They are however mechanically hard materials and so difficult to machine. For this reason they are generally formed into a simple **shape** by casting (pouring

Table 5.3 Magnetic properties of permanent magnet materials

Material	B_r T	H_c kA m^{-1}	$(BH)_{max}$ T A m^{-1}
Fe–Co–Ni–Al alloys			
Alni	0.56	46	10
Alnico	0.72	45	14
Alcomax 2	1.30	46	43
Columax	1.35	59	60
Hycomax 1	0.90	66	26
Ferrites			
Feroba 1	0.22	135	8
Feroba 2	0.39	150	29
Feroba 3	0.37	240	26
Precipitation hardening alloys			
Comalloy	1.0	20	10
Vicalloy 1	0.9	24	8
Cunife 1	0.57	36	12
Rare earth			
Samarium cobalt	0.9	670	160
Neodymium iron boron	1.1	890	220
Steels			
6% tungsten	1.05	5.2	2.4
6% chromium	0.95	5.2	2.4
3% cobalt	0.72	10	2.8
35% cobalt	0.90	20	7.5

the hot liquid metal into a mould) or sintering (pressing a powder into the form before heating to bond the particles together). Such materials are widely used for moving-coil instruments, loudspeakers and relays.

Ferrites have high coercivity values but only moderately high values of remanence and $(BH)_{max}$. They have a Curie temperature of the order of 450 °C and so are limited in the temperatures to which they can be used. They have high electrical resistivities, low density, are relatively cheap to manufacture and have a low cost per unit of available magnetic energy. They are ceramics and are formed by mixing together the appropriate compounds and then heating to form particles. These are then pressed into a die to the required shape and then heated. Such materials are widely used for loudspeakers, the field magnets in small d.c. motors, and holding catches for doors.

The rare earth alloys have high remanence, high coercivity and high values for $(BH)_{max}$. The neodymium iron boron alloy however suffers from the disadvantage that its magnetic properties show a rapid change with temperature.

Example 6

What will be the minimum volume of a tungsten steel magnet which is required to maintain a flux density of 0.1 T across an air gap 12 mm long and with a cross-sectional area of 400 mm^2? The leakage factor k_1 may be taken as 4.0 and the correction factor k_2 1.2. The value of $(BH)_{max}$ for the tungsten steel is 2.4 T A m^{-1}.

Answer

Using the equation [10] developed above

$$V_m = \frac{k_1 k_2 B_g^2 V_g}{B_m H_m \mu_0}$$

$$= \frac{4.0 \times 1.2 \times 0.1^2 \times 0.012 \times 400 \times 10^{-6}}{2.4 \times 4\pi \times 10^{-7}}$$

$$= 0.076 \text{ m}^3$$

Meter magnetic materials

The basic design of a *moving-coil galvanometer* is of a coil in a magnetic field which is always at right-angles to the sides of the coil. When a current flows through the coil it is acted on by a torque which causes it to rotate against the opposing torque provided by springs. See Chapter 4 for a more detailed discussion.

Figure 5.12(a) shows the traditional design of the magnetic circuit for such an instrument. The shaped pole pieces and the central cylinder of a soft magnetic material provide a radial field which is always at right-angles to the sides of the coil.

Fig. 5.12 Basic design of magnetic circuits in moving-coil galvanometers

(a)

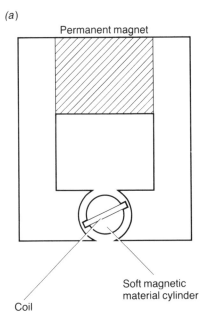

(b)

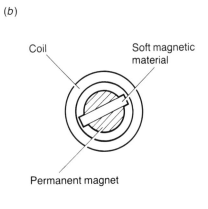

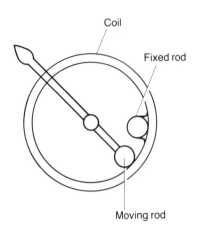

Fig. 5.13 Moving-iron galvanometer

That design however suffers from the disadvantages of being a magnetic circuit with a high leakage factor (k_1) and also occupies a large space. The high leakage factor occurs because the permanent-magnet material is some distance from the gap in which the field is required. A better design is that of Fig. 5.12(*b*). This has a central permanent-magnet cylinder, diametrically magnetised, inside a hollow cylinder of soft magnetic material. This still gives a radial field but with less leakage and a more compact form. The permanent-magnet material is likely to be a Fe–Co–Ni–Al alloy such as Hycomax.

The *moving-iron galvanometer*, in one form (Fig. 5.13), involves two soft magnetic material rods located inside a coil. When a current flows through the coil the rods become magnetised and repel each other, so causing a pointer to move across a scale. The instrument can be used for both direct and alternating currents. The material of the rods is required to have:

1 a high permeability so that large flux densities are produced and hence large repulsive forces between the rods;
2 low remanence, the rods should not remain magnetised when there is no current through the coil;
3 low coercivity, so that the rods are easily demagnetised;
4 low hysteresis losses, so that the rods do not become hot and change the temperature and hence resistance of the galvanometer coil.

This is a specification for a soft magnetic material such as Mumetal, a Fe–76% Ni alloy with a minimum relative permeability of about 300 000, a remanence of 1.4 T and a coercivity of 4 A/m.

Transformer core material

In considering the selection of a material for the core of a transformer the following points have to be considered:

1 a high permeability is required, so that high flux is produced with as little current as possible;
2 a high saturation flux density, since the maximum flux is limited by this;
3 hysteresis losses have to be kept to a minimum;
4 eddy currents in the core must be kept as low as possible since such currents circulating in the core result in heating of the core;
5 the material selected must be one that can be readily machined to the required shape.

The requirement for minimum hysteresis losses means that a soft magnetic material is required. To keep eddy currents low the core can be made up of thin sheets, electrically insulated from each other. This increases the electrical resistance and so reduces the size of the currents. In addition the eddy currents can be kept low by using a material with a high resistivity. A requirement for the machining of the core is that the material should not be brittle but have reasonable ductility. Iron–silicon alloys are able to fit the above requirements.

The hysteresis loss of iron is reduced by increasing the amount of silicon alloyed with it, the inclusion of 4% silicon reducing the hysteresis loss by over 50%. The resistivity of an iron–silicon alloy increases up to about 11% silicon and then decreases. At room temperature iron–silicon alloys become brittle with more than about 4% silicon. This becomes the limiting factor. Thus the optimum iron–silicon alloy for a transformer core is likely to have a maximum of about 4% silicon. An alloy with about 3% silicon is widely used. Such a steel is likely to have, when the crystals in the solid are not orientated in any particular direction, a maximum relative permeability of about 8000, total losses (hysteresis plus eddy current) which amount to about 3 W per kilogram of steel at a frequency of 50 Hz, and an electrical resistivity of about 48×10^{-8} Ω m. A steel with the crystals orientated all in the same direction as a result of rolling of the steel strip will have, in the direction of the crystals, a relative permeability of about 90 000, total losses per kilogram of about 1 W at 50 Hz and an electrical resistivity of about 48×10^{-8} Ω m. Both steels have a Curie temperature of about 745 °C.

High-frequency inductor core material

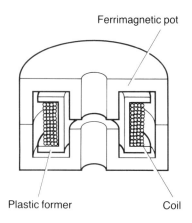

Ferrimagnetic pot

Plastic former   Coil

Fig. 5.14 High-frequency inductor

A prime requirement of the material used for the core of an inductor to be used at high frequencies is that it should have low losses. The small signals used with such inductors mean small flux changes in the core and hence hysteresis losses are generally negligible. The main losses arise from eddy currents because the high frequency means a high rate of change of flux in the core and so high induced e.m.f.s. The current in the core, and hence the eddy current losses, will depend on the electrical resistance of the core. The higher the resistance the smaller the current and so the smaller the losses. There is thus a requirement for a high-resistivity magnetic material for the core. For this reason ferrites are used, the ferrites generally being manganese–zinc or nickel–zinc ferrites (see earlier in this chapter and Table 5.2). Figure 5.14 shows the basic form of such an inductor. The coil is wound on a plastic former which is enclosed in a cylindrical ferrite pot. Because of the high resistivity of the ferrite there is generally no need to laminate it in order to reduce eddy currents.

Problems

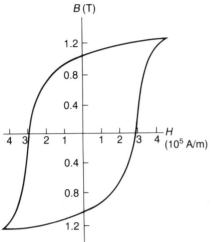

Fig. 5.15 Problem 5

1   What is (*a*) the intensity of magnetisation, (*b*) the flux density, produced when a magnetic field of strength 50 A/m is applied to a Fe–3% Si alloy with a relative permeability of 7000?

2   State the differences in magnetic properties between diamagnetic, paramagnetic and ferromagnetic materials.

3   Explain why iron is more strongly magnetic than copper.

4   Explain the difference between ferromagnetic and antiferromagnetic materials, considering the example of cobalt and manganese.

5   Figure 5.15 shows the hysteresis loop for a ferromagnetic material. Estimate from the graph the values of (*a*) the remanence, (*b*) the coercive field, (*c*) the saturation flux density, (*d*) the energy loss due to hysteresis during one cycle, (*e*) $(BH)_{max}$.

6   What is the main property difference between ferrites and ferromagnetic materials and what is responsible for the difference?

7   What will be the minimum volume of an alnico magnet which is required to maintain a flux density of 0.5 T across an air gap 2.0 mm long and with a cross-sectional area of 100 mm²? The leakage factor k_1 may be taken as 3.0 and the correction factor k_2 1.2. The value of $(BH)_{max}$ for the alnico is 14 T A m⁻¹.

8   What are the properties required for the magnetic material used for (*a*) the armature of a relay, (*b*) the record or playback heads of a magnetic-tape recorder, (*c*) magnetic tape for a magnetic-tape recorder?

6 Electron ballistics

Introduction

The use of electric and magnetic fields to control the motion of electrons in a vacuum is an essential feature of electronic tubes such as the diode, cathode ray tube, photoelectric cell, and photomultiplier. This chapter is about electron emission, the forces acting on electrons under the action of electric or magnetic fields, their motion in such fields and the application of such in electronic tubes.

Electron emission

Within metals we can consider there to be free electrons, not attached to any particular metallic ion (see metallic bonds in Ch. 1). Many electronic devices rely on getting some of these electrons to escape from the metal. Within the metal the free electrons move about between the positive metal ions but are held within the metal by the electric forces of attraction to the positive ions. However if energy is supplied to these electrons, as a result perhaps of heat being supplied to the material, some of the electrons may acquire sufficient energy for them to overcome the attractive forces of the ions and escape. The term *work function* ϕ is used to specify the energy needed to remove an electron from a material at absolute zero (see Ch. 7 for a discussion of the energies of electrons in metals). The reason for specifying absolute zero is that at any other temperature the electrons already will have received some energy and thus the amount that needs to be supplied for the electrons to escape will depend on the temperature. There are four methods of supplying the energy to the electrons to enable them to escape – heat resulting in what is termed thermionic emission, electromagnetic waves in what is termed photoelectric emission, kinetic energy of other electrons or particles bombarding the surface in what is termed secondary emission and by a high external electric field causing what is termed field emission.

With *thermionic emission* heat is used to cause electrons to be emitted from metal surfaces. When all the escaping electrons are collected the current per unit area of heated surface, i.e., the current density J, is given by the so-called *Richardson–Dushman equation* as

current per unit area $J = AT^2 e^{-\phi/kT}$ [1]

where A is a constant, ϕ the work function, k Boltzmann's constant and T is the temperature on the kelvin scale. In electronic tubes the three forms of emitting material used are tungsten, thoriated tungsten and oxides of barium and strontium on a nickel base. Table 6.1 shows the basic characteristics of such emitters. A low work function permits a low working temperature and consequently longer life for the emitter.

Table 6.1 Thermionic emitter materials

Material	Work function eV	Working temp. K	Current density 10^3 A/m^2
Tungsten	4.5	2500	2.5
Thoriated tungsten	2.6	1900	15
Barium + strontium oxides on nickel	1.0	1100	10

Note The current density quoted is that at the working temperature. The energy unit used for the work function is the electron volt (eV). This is the energy given to an electron when it is accelerated through a potential difference of 1 V and is 1.6×10^{-19} J (see Ch. 2).

Photoelectric emission is the emission of electrons from a metal as a result of it being illuminated by electromagnetic radiation, such as visible light, ultraviolet or infra-red radiation. Electromagnetic radiation can be considered to be propagated in the form of small packets of energy, called *photons*. Emission occurs if the energy supplied to electrons by the photons colliding with them is sufficient to give some electrons enough energy to escape. The energy E of a photon is proportional to the frequency f of the radiation, thus the higher the frequency the greater the energy.

$E = hf$ [2]

where h is a constant called Planck's constant. High frequency means short wavelengths, since wave speed c is frequency multiplied by the wavelength λ. This means that the short-wavelength ultraviolet photons have more energy than those of visible light and these in turn have more energy than those of the longer-wavelength infra-red radiation. Illuminating a surface is thus a bombardment of that surface by photons. To 'dig' an electron out of a material a certain energy is required, this being the work function ϕ. The energy supplied by a photon giving all its energy to an electron in the material must be greater than the work function if the electron is to escape.

$$\text{Maximum escaped electron energy} = hf - \phi \qquad [3]$$

The lowest frequency, for a particular surface, for which emission can occur is when $hf = \phi$. For some surfaces the value of their work function is high enough to mean that visible light has insufficient energy and only ultraviolet radiation can cause emission. For visible light to cause emission the work function has to be less than about 2 eV (i.e., 3.2×10^{-19} J, see Ch.2 for a discussion of the electron volt unit of energy). Caesium and composite materials involving caesium have a low work function and are thus used for photoelectric cells which have to respond to visible light.

Changing the frequency or wavelength of the radiation only affects the energy of the escaping electrons, it has no effect on the numbers escaping. For more to escape more photons must bombard the surface. This means increasing the intensity of illumination. Higher intensity means more photons.

With *secondary emission* electrons, or other particles, colliding with a material may have sufficient energy to give electrons in the surface of the material enough energy to escape. Thus a single electron incident on a surface might result in the escape of one or more secondary electrons and thus give rise to a multiplication of the number of free electrons. The ratio of the number of secondary electrons to the number of primary electrons is called the *secondary emission yield*.

With *field emission* the presence of a high electric field near a metallic surface can, at room temperature, pull some of the free electrons out of the surface.

Example 1

If the temperature of a tungsten filament is raised from 2000 K to 2500 K by what factor will the current density obtained from the filament be increased?

The work function of tungsten = 4.5 eV
1 eV = 1.6×10^{-19} J
Boltzmann's constant $k = 1.38 \times 10^{-23}$ J/K

Answer

According to the Richardson–Dushman equation [1]

$$J = AT^2 e^{-\phi/kT}$$

Thus

$$\frac{J_1}{J_2} = \frac{T_1{}^2 \exp}{T_2{}^2 \exp} \left(\frac{-\phi/kT_1}{-\phi/kT_2}\right)$$

$$= \frac{2500^2 \exp}{2000^2 \exp} \left[\left(\frac{-4.5 \times 1.6 \times 10^{-19})/(1.38 \times 10^{-23} \times 2500}{-4.5 \times 1.6 \times 10^{-19})/(1.38 \times 10^{-23} \times 2000}\right)\right]$$

$$= 288$$

Example 2

Caesium has a work function of 2.1 eV. What is the maximum wavelength of electromagnetic radiation which will result in the photoelectric emission of electrons?

Planck's constant $h = 6.6 \times 10^{-34}$ J s
1 eV $= 1.6 \times 10^{-19}$ J
Speed of light $= 3.0 \times 10^8$ m/s

Answer

The minimum frequency f for emission to occur is when

$$hf = \phi$$

where ϕ is the work function. Since the speed of light $c = f\lambda$, the maximum wavelength λ is when

$$\frac{hc}{\lambda} = \phi$$

$$\lambda = \frac{hc}{\phi} = \frac{6.6 \times 10^{-34} \times 3.0 \times 10^8}{2.1 \times 1.6 \times 10^{-19}} = 5.9 \times 10^{-7} \text{ m} = 590 \text{ nm}$$

This wavelength corresponds to the yellow part of the visible spectrum. Thus only shorter wavelengths, i.e., more blue, will cause emission of electrons.

Example 3

What is the maximum velocity of electrons emitted from a surface with a work function of 2.0 eV when it is illuminated by light of wavelength 500 nm?

1 eV $= 1.6 \times 10^{-19}$ J
$h = 6.6 \times 10^{-34}$ J s
Speed of light $= 3.0 \times 10^8$ m/s
Mass of an electron $= 9.1 \times 10^{-31}$ kg

Answer

As indicated in equation [3]

maximum escaped electron energy $= hf - \phi$

Since $c = f\lambda$, then

maximum escaped electron energy $= \dfrac{hc}{\lambda} - \phi$

Thus since the kinetic energy of the escaped electron is $\frac{1}{2}mv^2$, then

$$v^2 = \frac{2}{m}\left(\frac{hc}{\lambda} - \phi\right)$$

$$= \frac{2}{9.1 \times 10^{-31}}\left(\frac{6.6 \times 10^{-34} \times 3.0 \times 10^8}{500 \times 10^{-9}} - 2.0 \times 1.6 \times 10^{-19}\right)$$

$$v = 4.1 \times 10^5 \text{ m/s}$$

Force on electrons in electric fields

An electron, with charge q, in an electric field of strength E will experience a force F (see Ch. 2, equation [14]) where

$$F = qE \tag{4}$$

The direction of the force will be in the opposite direction to that of the electric field since the charge on the electron is negative and the field direction gives the direction of the force on a positive charge.

The force will cause the electron to accelerate in the direction of the force. Thus using Newton's second law of motion, the acceleration a will be given by

$$F = ma = qE \tag{5}$$

where m is the mass of the electron.

The work done in accelerating the electrons through a distance s in the direction of the force is

$$\text{work done} = Fs$$

and hence substituting for F using equation [5]

$$\text{work done} = qEs$$

The electrons in accelerating gain in velocity and hence kinetic energy. Thus the change in kinetic energy as a result of being accelerated by an electric field is

$$\text{change in kinetic energy} = \tfrac{1}{2}mv^2 - \tfrac{1}{2}mu^2 = \text{work done}$$

and so

$$\text{change in kinetic energy} = qEs$$

But the electric field is the potential gradient, thus if the potential changes by V in the distance s then the electric field is V/s and so

$$\text{change in kinetic energy} = qV \tag{6}$$

Example 4

What is the force acting on an electron when it passes into the electric field between two parallel conducting plates, 10 mm apart, if there is a potential difference of 1000 V between the plates?

Charge on the electron $= 1.6 \times 10^{-19}$ C

Answer

The force on an electron is given by equation [4] as

$$F = qE = \frac{qV}{d} = -\frac{1.6 \times 10^{-19} \times 1000}{0.010} = -1.6 \times 10^{-14} \text{ N}$$

Example 5

In a thermionic diode with plane parallel electrodes electrons are emitted from the cathode with negligible velocities and accelerated to the anode by a potential difference of 100 V between the anode and cathode. What will be the velocity of the electrons when they reach the anode?
> Charge on electron = 1.6×10^{-19} C
> Mass of electron = 9.1×10^{-31} kg

Answer

Using equation [6]

$$Vq = \tfrac{1}{2}mv^2$$

$$v^2 = \frac{2Vq}{m}$$

Hence

$$v^2 = \frac{2 \times 100 \times 10^{-19}}{9.1 \times 10^{-31}}$$

$$v = 4.7 \times 10^6 \text{ m/s}$$

Electron gun

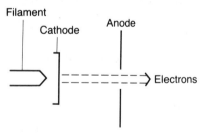

Fig. 6.1 Electron gun

The electron gun assembly (Fig. 6.1) used in cathode ray tubes uses a filament to heat a coated cathode and so produce electrons, as a result of thermionic emission. The electrons are then accelerated through a potential difference to an anode, which is positive with respect to the cathode. The anode contains a central hole so that the accelerated electrons then pass through the anode and are able to form a beam of electrons. The velocity v of the electrons in the beam is given by equation [6], when the electrons are emitted from the cathode with negligible velocities, as

$$\tfrac{1}{2}mv^2 = qV$$

where m is the mass of an electron, q its charge, V the potential difference between the anode and cathode.

Example 6

An electron gun assembly has a potential difference between the anode and cathode of 300 V. What will be the velocity of the electrons on leaving the anode if they can be considered to have a negligible velocity on emission from the cathode?

Charge on electron $= 1.6 \times 10^{-19}$ C
Mass of electron $= 9.1 \times 10^{-31}$ kg

Answer

Using equation [6]

$$\tfrac{1}{2}mv^2 = Vq$$

$$v^2 = \frac{2 \times 300 \times 1.6 \times 10^{-19}}{9.1 \times 10^{-31}}$$

$$v = 1.0 \times 10^7 \text{ m/s}$$

Deflection of electron beam in an electric field

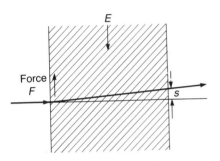

Fig. 6.2 Electron moving into an electric field

If electrons are moving with some velocity v which is at right-angles to the field direction then the force of qE will act on each electron in a direction which is at right-angles to its initial motion. This force will produce an acceleration in the direction of the force but since there is no component of the force in the direction of its initial motion then there is no acceleration in that direction and hence no change in the velocity in that direction. There is, however, an acceleration in a direction at right-angles to its initial motion. Thus the electron has two components to its motion, a constant velocity in the initial direction and an acceleration at right-angles to this direction, as in Fig. 6.2.

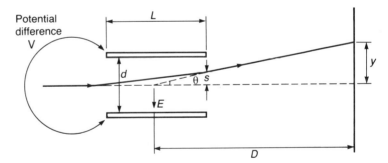

Fig. 6.3 Electron beam deflected by a potential difference

Consider an electron moving initially with a velocity v at right-angles to the electric field produced by a pair of parallel plates between which a potential difference V is maintained (Fig. 6.3). The field is uniform and so the acceleration is constant and the distance s moved along the line of action of the field is given by

$$s = \tfrac{1}{2}at^2 = \tfrac{1}{2}(qE/m)t^2 \qquad [7]$$

The time t the electron is between the plates depends on its initial velocity v and the length L of the plates. Since its velocity at right-angles to the electric field does not change then the time between the plates is

$$t = \frac{L}{v}$$

Thus

$$s = \frac{1}{2}(qE/m)(L/v)^2$$

The electron thus suffers a sideways deflection s which is proportional to the electric field strength. Thus for a beam of electrons passing through a pair of parallel conducting plates, the deflection of the beam is proportional to the electric field strength between the plates. But the electric field strength is the potential gradient (V/d), hence the deflection s is proportional to the potential difference between the plates.

$$s = \frac{1}{2}(qV/md)(L/v)^2$$

Once the electron has left the electric field between the plates there are no further forces acting on it so it continues in motion in a straight line. Thus if some distance D away it hits a screen then the deflection of the beam on the screen of y is given, to a reasonable approximation (the path between the plates is parabolic but reasonably approximates to a straight line), by

$$\tan \theta = \frac{s}{\frac{1}{2}L} = \frac{y}{D}$$

$$y = \frac{2Ds}{L}$$

Thus

$$y = \frac{DqVL}{mdv^2} \qquad [8]$$

This discussion has assumed that the field is perfectly uniform between the plates and changes abruptly from this constant value to zero at the edges of the plates. This is obviously not the case in reality but the above is a reasonable approximation.

The deflection of a beam of electrons by a potential difference applied between two parallel plates is the basis of the deflection system used with cathode ray oscilloscopes. This has two sets of parallel plates at right-angles to each other. One pair is used to produce a deflection in the x-direction and the other in the y-direction.

Example 7

A cathode ray tube has electrostatic deflection plates of length 20 mm. If a beam of electrons is accelerated through a potential difference of 100 V before passing between the plates what will be the deflection of the beam of electrons on a screen 200 mm from the plates when a potential difference of 20 V is applied to the plates? The plates have a separation of 10 mm.

Answer

The velocity v of the electrons on entering the electric field between the plates is given by equation [6]

$$\tfrac{1}{2}mv^2 = V_a q$$

where V_a is the potential difference used to initially accelerate the electrons. Thus using equation [8]

$$y = \frac{DqVL}{mdv^2} = \frac{DVL}{2dV_a}$$

$$= \frac{200 \times 10^{-3} \times 20 \times 20 \times 10^3}{2 \times 10 \times 10^{-3} \times 100}$$

$$= 0.040 \text{ m} = 40 \text{ mm}$$

An electrostatic lens

Electrostatic lenses are used in cathode ray tubes to take the electrons that are emitted from the cathode at different initial angles and bring them all to the same spot on the screen. The term lens is used because it is doing a similar job to a lens focusing light.

Figure 6.4 shows an example of an electrostatic lens. It depends for its action on the electric fields between adjacent cylinders. The electric fields are not uniform, the figure showing the electric lines of force. When the electrons arrive in such a field they experience forces which act in the direction of the field. These forces can be resolved into two components, one along the axis of the cylinder and one at right-angles to it. Thus at point A the electrons experience a force which deflects the beam towards the tube axis and a force along the axis of the tube which speeds up the electrons. At point B the electrons experience a force which deflects the beam away from the tube axis. Though the size of the force is the same as at A the deflection is less because the electrons are moving faster at this point and spend less time in this part of the field (see eqn [7]). Thus the net result of the electrons passing through this region between the first two cylinders is that the electron beam receives a net deflection towards the axis of the cylinders. When the electrons reach point C they experience a force which deflects the beam away from the axis and a force along the axis which slows the electrons down. At point D the force deflects the beam towards the axis, but because the electrons have been slowed down they spend longer in the region of the field and so the deflection towards the axis is greater than that away from the axis at C. The net result is a deflection towards the axis. Because of the shape of the electric fields between the cylinders the electrons moving in lines further out from the axis experience greater deflections than those moving nearer to the axis, indeed those moving along the axis experience no deflecting force. The net result is that the beam of electrons is focused.

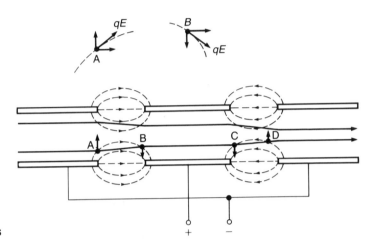

Fig. 6.4 Electrostatic lens

Force on electrons in a magnetic field

For a conductor of length L in a magnetic field of flux density B at right-angles to the conductor then the force F acting on it when it is carrying a current I is (see Ch. 4, eqn [42])

$$F = BIL$$

with the force, current and flux density all mutually at right-angles to each other. But current is just the rate of movement of charge and thus for an electron, charge q, moving with a velocity such that it covers a distance L in time t then

$$I = \frac{q}{t}$$

but if the velocity of the electron is v then $v = L/t$ and so

$$I = \frac{qv}{L}$$

Thus for an electron moving with a velocity v at right-angles to a magnetic field of flux density B the force acting on it is

$$F = Bqv \qquad [9]$$

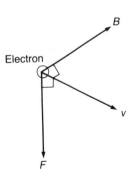

Fig. 6.5 Force on an electron moving in a magnetic field

The direction of this force is at right-angles to both the direction of motion of the electron and the magnetic field (Fig. 6.5). Since this force is at right-angles to the motion of the electron it has no effect on the size of v but can only change the direction. Because the force is always at right-angles to the motion the electron is made to move, within the magnetic field, in a circular path. The radius r of such a path when there is a centripetal force F is given by

$$F = \frac{mv^2}{r}$$

where m is the mass of the electron. Thus

$$Bqv = \frac{mv^2}{r}$$

$$r = \frac{mv}{Bq} \qquad [10]$$

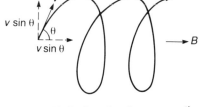

Fig. 6.6   Helical motion in a magnetic field

When the magnetic field is acting at some angle θ to the direction of motion of an electron then the electron velocity v can be resolved into two components, $v \cos \theta$ in the direction of the field and $v \sin \theta$ at right-angles to the field (Fig. 6.6). There is only a force acting on an electron due to the magnetic field when the magnetic field is at right-angles to the motion. Thus there will be no force acting in the $v \cos \theta$ direction and so the velocity component in this direction will be unchanged. The $v \sin \theta$ component is, however, at right-angles to the field and so there will be a force acting which will lead to circular motion of radius (using equation [10]) $(mv \sin \theta)/Bq$. Thus the motion of the electron is a circular motion superimposed on the $v \cos \theta$ motion, i.e., a helical motion.

Example 8

What is the radius of the circular path of an electron beam in a uniform magnetic field of flux density 4.0×10^{-6} T when moving at right-angles to it with a velocity of 1.0×10^5 m/s?
  Charge on the electron = 1.6×10^{-19} C
  Mass of the electron = 9.1×10^{-31} kg

Answer

Using equation [10]

$$r = \frac{mv}{Bq} = \frac{9.1 \times 10^{-31} \times 1.0 \times 10^5}{4.0 \times 10^{-6} \times 1.6 \times 10^{-19}} = 0.14 \text{ m}$$

Deflection of electron beam in a magnetic field

Magnetic deflection of electron beams is widely used in the cathode ray tube of TV receivers. The magnetic field is produced by a pair of coils outside the tube. A current through the coils produces a magnetic field which is at right-angles to the electron beam and hence results in a force acting on the

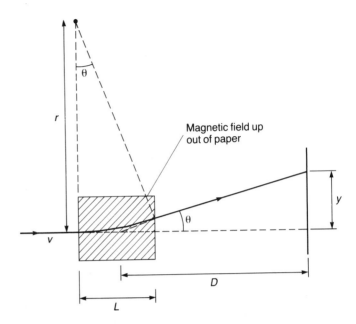

Fig. 6.7 Electron beam deflected by a magnetic field

electrons and producing a deflection of the beam (Fig. 6.7). If the field can be assumed to be constant over a length L and zero outside this length then the electrons will move in a circular path of radius r within the field and in straight lines outside the field. If at a distance D from the centre of the magnetic field the deflection is y then

$$\tan \theta = \frac{y}{D}$$

and if the angle of deflection θ is small then $\tan \theta$ approximates to θ which is approximately L/r, where r is the radius of the electron beam in the magnetic field. Thus

$$\frac{y}{D} = \frac{L}{r}$$

and since, from equation [10], $r = mv/Bq$, then

$$y = \frac{LDBq}{mv} \qquad [11]$$

The deflection y is thus proportional to the flux density B and since this is proportional to the current through the coils, then the deflection is proportional to the current.

If the beam of electrons had been accelerated through a potential difference of V_a before entering the magnetic field then the velocity v of the electrons would be given by

$$\tfrac{1}{2}mv^2 = V_a q$$

and so substituting for v in equation [11]

$$y = \frac{LDB\sqrt{q}}{\sqrt{(2mV_a)}} \qquad [12]$$

A magnetic lens

Consider a beam of electrons travelling along the axis of a coil (Fig. 6.8). If the electrons in the beam have velocities at different angles to the axis then the beam will spread out. However if a magnetic field is applied the electrons will be acted on by forces related to their velocity component at right-angles to the axis. Thus if an electron has a velocity v at angle θ to the axis then the velocity component at right-angles to the axis is $v \sin \theta$. Thus the force acting on it is, using equation [9],

$$F = Bqv \sin \theta$$

This force will cause the electron to move in a circular path of radius r, where

$$\frac{m(v \sin \theta)^2}{r} = Bqv \sin \theta$$

$$r = \frac{mv \sin \theta}{Bq}$$

The time T taken for the electron to complete one orbit of this circular path is

$$T = \frac{2\pi r}{v \sin \theta}$$

$$T = \frac{2\pi m}{Bq} \qquad [13]$$

The time taken to complete an orbit is thus independent of the velocity of the electron and its direction. All the electrons in the beam, regardless of their directions of motion, will thus take the same time to complete one orbit.

The electrons all have a velocity component along the axis of the coil and thus the motion of each electron is helical. If all

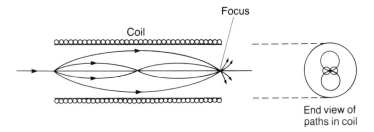

Fig. 6.8 Magnetic lens

the electrons have essentially the same velocity along the axis v_a then a consequence of them all taking the same time to complete an orbit is that after that time they will all converge to the same point on the coil axis. This will be a distance of v_aT from the start of the magnetic field, i.e., using equation [13]

$$\text{distance} = \frac{2\pi m v_a}{Bq} \qquad [14]$$

The electron beam has thus been focused. The above thus describes a magnetic lens.

Example 9

What is the axial magnetic field required to focus electrons, which have been accelerated through 1.0 kV, to a point in a distance of 100 mm?
Charge on electron = 1.6×10^{-19} C
Mass of electron = 9.1×10^{-31} kg

Answer

Using equation [6]

$$\tfrac{1}{2}mv^2 = Vq$$

and equation [14]

$$\text{distance} = \frac{2\pi mv}{Bq} = \frac{2\pi \sqrt{(2mV)}}{B\sqrt{q}}$$

$$B = \frac{2\pi \times \sqrt{(2 \times 9.1 \times 10^{-31} \times 1000)}}{0.100 \times \sqrt{(1.6 \times 10^{-19})}} = 6.7 \times 10^{-3} \text{ T}$$

Vacuum electronic tubes

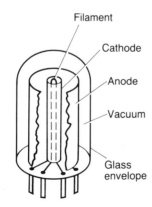

Fig. 6.9   Thermionic diode

The *thermionic diode* (Fig. 6.9) consists of a central filament which is used to heat the cathode and an anode to collect electrons. As a result of thermionic emission electrons are produced from the cathode. Because so many electrons are emitted from the cathode a cloud of negative charge, known as a *space charge*, forms round the cathode. Electrons from this space charge are attracted to an anode when it is at a positive potential with respect to the cathode. The diode is a rectifying device because the electron flow can only be in one direction, from cathode to anode, and thus only when the anode is at a positive potential with respect to the cathode.

The *vacuum photoemissive cell* (Fig. 6.10) consists of a metal half cylinder cathode which is coated on its inside with a photoemissive material and a central rod anode. When electromagnetic waves, of suitable frequency, are incident on

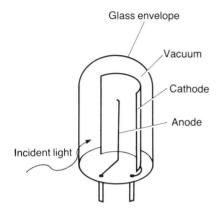

Fig. 6.10 Vacuum photoemissive cell

the photoemissive material electrons are emitted. When the anode is at a positive potential with respect to the cathode electrons move from cathode to anode and a current is produced which is related to the intensity of light falling on the photoemissive material.

The *photomultiplier* (Fig. 6.11) is a combination of a photoemissive cathode and a series of electrodes called dynodes. The electrons produced at the cathode are focused and accelerated to the first dynode by a suitable potential difference and strike the dynode with sufficient energy to cause secondary emission. As a consequence the number of electrons is increased. These are then focused and accelerated to the second dynode where the number is still further increased by secondary emission. If the number of electrons is multiplied by f at each dynode, i.e., the secondary emission

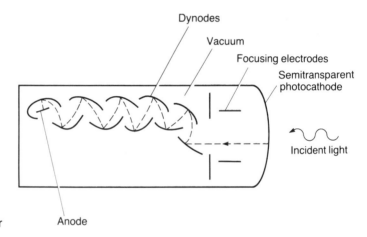

Fig. 6.11 Photomultiplier

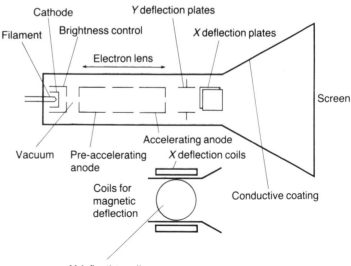

Fig.6.12 Cathode ray tube

yield is f, and there are n dynodes then the number of electrons produced at the cathode is multiplied by f^n by the photomultiplier. The value of f depends on the material on the surface of the dynodes and is typically of the order of 5 to 10 but can be as high as 50. The number of dynodes is generally between 8 and 14. As a consequence there is high internal amplification, of the order of 10^4 to 10^8 and the photomultiplier can be used to detect very low levels of light.

The *cathode ray tube* (Fig. 6.12) consists of a source of electrons, acceleration and focusing of the electron beam, means of producing deflections of the beam in the X and Y directions, and a fluorescent screen which shows the position of the electron beam. The electrons are produced by a current through a filament being used to heat a cathode and so give thermionic emission. The electrons are accelerated to the first anode through a hole in an electrode. By controlling the potential of this electrode, typically between 0 and 100 V negative with respect to the cathode, so the numbers of electrons forming the electron beam can be controlled and hence the brightness of the impact of the electrons at the screen. The basic cathode ray tube is likely to have for acceleration and focusing an arrangement of cylindrical electrodes of the form shown in the figure with the adjustment of the potential of the central electrode being used to adjust the focusing (see earlier in this chapter and Fig. 6.3 for an explanation of this form of electron lens). Typically the final anode will be about 3 kV to 4 kV positive with respect to the cathode. The deflection system may be by electric or magnetic fields. For electric fields there are two pairs of parallel plates, a potential difference between the plates being used to produce a deflection. For magnetic field deflection two pairs of coils are used, a current through a pair of coils producing a magnetic field and hence deflection of the beam. Deflection by means of an electric field is commonly used in cathode ray oscilloscopes, while magnetic deflection, because it is more easily able to produce large deflections (though it is slower), is used in the cathode ray tubes used for televisions, alphanumeric and graphical displays. The impact of the electrons on the phosphor-coated screen results in light being given off by the phosphor. The amount of light given off depends on the type of phosphor used and the velocity of the electrons. For an oscilloscope to respond to a fast-occurring event and have a high writing speed a high-velocity beam of electrons is required. While this could be achieved by increasing the potential difference between the anodes and the cathode this leads to the problem that it is much more difficult to deflect the electron beam. For deflection by electric fields much higher fields are required. This problem can be overcome by carrying out the acceleration of the electrons in two stages, prior to the deflection system and after it. Such a tube is known as a postdeflection acceleration tube. The electrons in bombarding the phosphor release secondary emission electrons. These are collected by a conductive coating on the inside of the glass tube which is connected to the second anode.

Example 10

A photomultiplier has eight dynodes with each having a secondary emission yield of 10. What is the internal amplification of the photomultiplier?

Answer

The internal amplification is f^n, where f is the secondary emission yield and n the number of dynodes. Thus the amplification is 10^8.

Problems

Charge on the electron $= 1.6 \times 10^{-19}$ C
Mass of the electron $= 9.1 \times 10^{-31}$ kg
Planck's constant $h = 6.6 \times 10^{-34}$ J s
Boltzmann's constant $k = 1.38 \times 10^{-23}$ J/K
Speed of light $= 3.0 \times 10^8$ m/s
1 eV $= 1.6 \times 10^{-19}$ J

1 Briefly discuss the various methods which can be used to cause electrons to be emitted from surfaces.
2 A photoelectric emissive cell has a photocathode with a work function of 1.9 eV. What will be (*a*) the minimum frequency of light for which emission will occur, (*b*) the maximum wavelength of light for which emission will occur, (*c*) the maximum kinetic energy of the electrons when the cell is illuminated by light of wavelength 500 nm, (*d*) the maximum velocity of the electrons when illuminated by this light?
3 Electrons emitted from a cathode are accelerated through a potential difference of 300 V, what will be their velocity after the acceleration?
4 An electron gun has a potential difference of 200 V between the anode and the cathode. If the electrons are emitted from the cathode with negligible velocity, what will be the velocity of the electrons on emerging from the anode?
5 What is the force acting on an electron when it moves between two parallel conducting plates if they are 10 mm apart and there is a potential difference of 500 V between them?
6 In a cathode ray tube a beam of electrons, after having been accelerated through 800 V, passes between a pair of parallel conducting plates of length 20 mm and separation 8 mm. The plates are 200 mm from the screen. What will be the deflection of the electron beam at the screen when a potential difference of 50 V is applied between the plates?
7 A beam of electrons after having been accelerated through 400 V passes into a magnetic field of 4.0 μ*T* which is at right-angles to

the path of the electrons. What will be the radius of the circular path of the electrons?

8   A beam of electrons is accelerated from rest through 1.0 kV before entering a magnetic field which is at right-angles to the beam. In the magnetic field the electron beam follows a circular path of radius 0.10 m, what is the magnetic flux density?

9   A beam of electrons enters a current-carrying solenoid at a small angle to the solenoid axis. What will be the form of the path of the beam through the solenoid if the magnetic field can be assumed to be constant within the solenoid?

10   What is the axial magnetic field required to focus electrons which have been accelerated through 2.0 kV to a point a distance of 60 mm along the axis?

11   An electron with a velocity of 1.0×10^7 m/s enters a magnetic field of 0.050 T at an angle of 20° to the field direction. What is the distance along the field direction at which the electron completes two turns of its helical motion?

12   A beam of electrons of velocity 2.0×10^6 m/s enters an electric field of 1.0×10^5 V/m at right-angles to its path. What magnetic field should be applied, over the same region and perpendicular to the electric field, for the electron beam to emerge from the field undeviated from its original path?

7 Electrical conductivity

In Chapter 1 an explanation of the high electrical conductivity of metals was given in terms of metals having free electrons which were able to move freely through the metal. Insulators were considered to have no free electrons. In this chapter this explanation is looked at in more detail with a consideration of the energy levels of electrons in atoms. This leads to a greater understanding of electrical conductivity and, in the next chapter, the electrical behaviour of junctions between materials and such electronic components as p–n junction diodes and transistors.

Energy levels in solids

The electrons in isolated atoms occupy discrete energy levels (see Ch. 1), these being determined as a consequence of each electron having its own unique set of quantum numbers with no two electrons having the same set. Thus, for example, for the 2s level the possible sets of quantum numbers indicate that two electrons can occur. However if the atoms are packed together to form a solid the atoms are no longer individual atoms. An electron is not then under the electrostatic influence of just its own nucleus but also all the other electrons and nuclei of all the atoms around it. As a consequence we have to consider the energy levels for the solid as a whole, bearing in mind the vital point that each electron has still to have its own set of quantum numbers. Thus if we take the simple case of two atoms coming together, each energy level splits into two. In the case of the 2s level which can accommodate two electrons for an individual atom, with the two atoms we need two 2s levels in order to accommodate the four electrons. This means that in the solid with N atoms, each level splits into N levels. Thus since for a typical crystal N is about 10^{29} atoms per cubic metre each level splits into 10^{29} levels. The separation between these levels becomes so small

Fig. 7.1 Energy levels for lithium, (a) isolated atoms, (b) as a solid

that we can regard each of the levels of the isolated atoms becoming continuous energy bands.

Figure 7.1 shows the energy levels for lithium as an isolated atom and as solid lithium. The isolated atom has two electrons in the 1s level (K shell) and one electron in the 2s level (L shell). The 1s level is full while the 2s level is only half occupied. Higher levels are unoccupied. When N lithium atoms are packed together to form the solid then the 1s band has $2N$ electrons in N energy levels and the 2s band N electrons in N energy levels. The effect of packing the atoms together to form the solid is to broaden the energy levels into bands, the 2s level being more broadened than the 1s level. The 1s band is full while the 2s band is only half occupied. Higher energy bands are unoccupied. With the lithium the electrons involved in it making bonds are the outer 2s electrons, the atom being said to have a valence of 1 since there is just one electron in this level. Thus in the case of the solid the 2s band is referred to as the valence band (note that for solid lithium the 2p band and the 2s band broaden sufficiently to merge and so the valence band is the combined band). In general, the *valence band* is the highest energy band containing occupied energy levels at a temperature of 0 K. The reason for specifying the temperature is that at higher

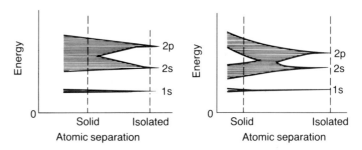

Fig. 7.2 Energy levels and atomic separation, (a) lithium, (b) carbon

temperatures electrons may have received sufficient energy to move to higher energy levels.

Figure 7.2 shows how energy levels broaden as the individual atoms are brought closer together. The closer the atoms the more the energy levels broaden, with the outer levels broadening more than the inner ones. In the case of some elements as the atomic separation is reduced the levels may broaden sufficiently for neighbouring levels to merge and form an even broader band. For some elements such a merged broad level may split into two separate bands at even closer atomic separations.

Band theory of solids

The highest energy band containing occupied energy levels at 0 K in a solid is called the *valence band*. For some elements the valence band is only partially filled with electrons while for others it is completely filled. The energy band immediately above the valence energy band and which contains vacant energy levels at 0 K is called the *conduction band*. In some cases there is no gap between the valence energy levels and the conduction band, in other cases there is a gap. The energy gap between the valence and conduction bands depends on the element concerned. Band theory for solids is concerned with these two bands.

When a potential difference is connected across a piece of material an electric field is produced in the material (see Ch.2). In the case of a good electrical conductor this potential difference produces a current, i.e., a movement of charge. In the case of a perfect insulator there is no current and hence no movement of charge. Electrical conduction in a solid requires there to be charges which are free to respond to the application of an electric field and gain energy. But the electrons in atoms can only exist at certain specified energy levels. This means that electrons must be able to move to an empty energy level. Electrons in a partially filled energy band have many empty energy levels they can move to. However, electrons in full energy bands are not able to respond to the electric field and contribute to the electrical conductivity unless they are able to jump across the gap into the next unfilled band.

Diamond, i.e., carbon, is an electrical insulator. In diamond the valence band is full and is separated from the conduction band by about 5 eV (the electron-volt is a unit of energy, being the energy needed to accelerate an electron through 1 V, i.e., 1.6×10^{-19} J) (Fig. 7.3(a)). This energy gap is too big for electrons to jump across it at normal temperatures. Thus there are no charge carriers for a current to occur. Some materials, such as germanium and silicon, have energy gaps

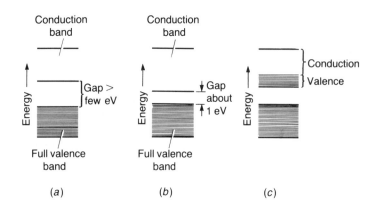

Fig. 7.3 Energy bands, (a) insulator, (b) semiconductor, (c) good conductor

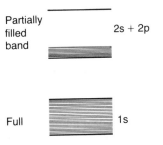

Fig. 7.4 Energy bands for lithium

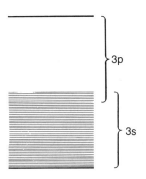

Fig. 7.5 Energy bands for magnesium

between the valence and conduction bands which are about 1 eV (Fig. 7.3(b)). This is a small enough gap for some electrons to be able to jump across it at room temperature. Typically at about room temperature germanium with an energy gap of 0.66 eV has about one atom in 10^9 for which electrons have jumped the gap, giving about 10^{19} 'free' electrons per cubic metre of material. Silicon has an energy gap of 1.12 eV and at room temperature about one atom in 10^{12} for which electrons have jumped the gap, giving about 10^{16} 'free' electrons per cubic metre of material. Such materials are known as *semiconductors*. Good conductors have only partially filled bands and so there is no gap between the valence energy levels and the conduction energy levels (Fig. 7.3(c)). Such materials have about one free electron per atom, giving about 10^{28} 'free' electrons per cubic metre of material. Table 7.1 shows the relationship that occurs between the electrical resistivity at about room temperature and the energy gap which has to be jumped for electrical conduction to occur.

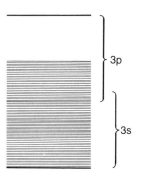

Fig. 7.6 Energy bands for aluminium

Table 7.1 Conductors, semiconductors and insulators

Material	Resistivity $\Omega\ m$	Energy gap eV	'Free' electrons per m^3
Good conductor	10^{-8} to 10^{-4}	0	10^{28}
Semiconductor	10^{-4} to 10^7	about 1	10^{16} to 10^{19}
Insulator	10^{12} to 10^{20}	> a few	0

Note The resistivity data refer to about room temperature.

Conductors

Lithium has two electrons in the 1s shell and one in the 2s shell. This isolated electron in an outer s shell is characteristic of elements which have the valence 1. For such elements the energy bands are of the form shown in Fig. 7.4, the 2s and 2p bands merging to form a large band which is only partially occupied (see also Fig. 7.2(*a*)). Lithium, and the other similar elements, are thus good conductors since an electric field is able easily to move electrons into vacant energy levels.

Magnesium has two electrons in the 1s shell, two in the 2s shell, six in the 2p shell and two in the 3s shell, having a valence of 2. Thus all its shells are full. Solid magnesium, like other valence 2 elements, is a good conductor of electricity. This is because the 3s and 3p energy bands merge (Fig. 7.5), and so the 3s electrons have the vacant 3p energy levels into which they can easily move when an electric field is applied.

Aluminium has two electrons in the 1s shell, two in the 2s shell, six in the 2p shell, two in the 3s shell and one in the 3p shell, having a valence of 3. The 3p shell is thus only partially occupied and so there are vacant energy levels into which electrons can easily move when an electric field is applied (Fig. 7.6). Aluminium is thus a good conductor.

All good conductors have no gap between the valence band and the conduction band.

Insulators

Insulators are materials which have a substantial gap, more than 4 or 5 eV, between their valence and conduction energy bands. Diamond, i.e., carbon, is an insulator. Carbon has two electrons in the 1s shell, two in the 2s shell and two in the 2p shell, having a valence of 4. This would suggest that there will be vacant energy levels in the 2p shell. However, in diamond the bonding between carbon atoms is by covalent bonds (see Ch. 1) with electrons being shared between neighbouring atoms so that each atom has a share in eight 2p electrons. All the electrons are very tightly held between the atoms by this covalent bonding and so none is free. The consequence of this bonding is that diamond has a full valence band with a substantial gap between it and the conduction band (see Fig. 7.2(*b*)).

Graphite, another form of carbon, is not an insulator like diamond. This is because all the electrons in the graphite structure are not locked up in covalent bonds (see Ch. 1) and there are some of them available for conduction. Because of this the resistivity of graphite is about 10^{-4} Ω m and it is just on the limit of what might be termed a good conductor.

Insulators on the energy band model described above would appear to have infinite resistivity. This is not the case. A full valence band separated by a large gap from the empty conduction band is a necessary but not sufficient condition for an insulator. This is because conduction by electrons is just one of the means by which a current can flow through a material. In crystalline materials the dominant method of conduction can be by the movement of some of its ions, in polymers it can be by ions which occur as impurities.

Semiconductors

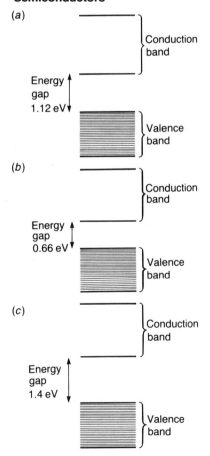

Fig. 7.7 Energy bands for (a) silicon, (b) germanium, (c) GaAs

Semiconductors are elements or compounds that in the solid state have an energy gap between the valence and conduction bands of less than about 4 eV. Germanium and silicon are examples of semiconductors. Both are in group IV of the Periodic table (see Ch. 1), and have the electronic structure of:

	1s	2s	2p	3s	3p	3d	4s	4p
Silicon	2	2	6	2	2			
Germanium	2	2	6	2	6	10	2	2

They have a valency of 4 and the atoms in the solid are held together by covalent bonds, each atom sharing electrons with each of four neighbours. They have a crystalline structure similar to that of diamond. All the electrons are thus locked up in bonds. Thus at 0 K there is a gap between a full valence band and the conduction band. However, unlike insulators such as diamond, the gap is relatively small and some of the valence electrons at room temperature have sufficient energy to jump the gap. This means that some of the electrons receive sufficient energy to break free from the bonds. Figure 7.7 shows the energy bands at 0 K for silicon and germanium.

In addition to the group IV elements, semiconductors have been produced as compounds between group III and group V elements, i.e., elements with three valence electrons forming covalent bonds with elements having five valence electrons. Examples of such compounds are gallium arsenide (GaAs), gallium phosphide (GaP), and indium antimonide (InSb). Such compounds have energy bands and gaps similar to those occurring with germanium and silicon. Figure 7.7(c) shows the energy bands for gallium arsenide at 0 K.

Example 1

Which of the following, at room temperature in the solid state, are conductors, semiconductors and insulators? The data given for each are the size of the energy gap between the valence and conduction bands.

Aluminium	0 eV
Aluminium phosphide	2.5 eV
Aluminium nitride	6.3 eV
Boron nitride	6 to 8 eV
Gallium antimonide	1.95 eV

Answer

Aluminium, because it has no gap, is a good conductor. Aluminium phosphide and gallium antimonide are semiconductors because the gap is fairly small, while aluminium nitride and boron nitride are insulators because the gap is more than about 4 or 5 eV.

Intrinsic and extrinsic semiconductors

In a semiconductor, such as germanium or silicon, the gap between the valence and conduction bands is about 1 eV. The energy supplied to the valence electrons by the material being at room temperature is enough for just a very small number of the electrons to be able to jump from the valence to the conduction band. However, when an electron jumps it leaves behind a vacant energy level, a hole in the valence band. Thus when an electric field is applied, not only can the electron in the conduction band move but also an electron in the valence band can move into the vacant valence band level. The vacancy in the valency band thus is shifted to another energy level. We talk of the hole in the valence level moving.

In terms of the bonds between, say, silicon atoms the situation is that each atom has four covalent bonds with neighbouring atoms, each bond being a shared pair of electrons. When a valence electron moves then one of the bonds is broken and there is a hole in the structure (Fig. 7.8). Under the action of an electric field an electron from a neighbouring atom can move into the hole and so the hole moves. We can picture the situation being rather like a line of people, a queue. When a gap appears near the front of the queue by someone moving out, the person behind the gap moves forward into it and the gap moves back one person. Then the next person moves forward into the gap. So as the people move forward into the gap the gap in the queue moves back down the queue. The gap moves in the opposite direction to the people. Thus when the electric field causes the electron in the conduction band to move in one direction, the hole apparently moves in the opposite direction. The hole thus moves in the same direction as a positive charged particle would. It is thus useful to think of holes as mobile positive

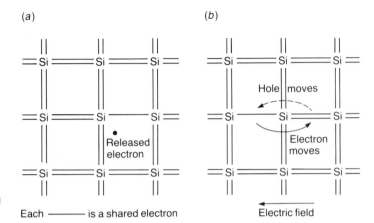

Fig. 7.8 (a) Release of a conduction electron, (b) movement of the resulting hole in an electric field

charges. Since all the holes are caused by movement of electrons from the valence band to the conduction band then there will be equal numbers of holes and conduction electrons. Both contribute to current flow through the material. Because there are equal numbers the semiconductor is said to be *intrinsic*. We can thus write

$$n = p = n_i$$

where n is the number of electrons per cubic metre, p the number of holes per cubic metre and n_i the number of either type of carrier per cubic metre, this being sometimes referred to as the *intrinsic carrier concentration* or density. The product of n and p is thus

$$np = n_i^2 \qquad \qquad [1]$$

Table 7.2 shows the numbers of conduction electrons and holes in intrinsic semiconductors.

This balance between the numbers of electrons and holes can be changed by replacing some of the atoms by atoms from other elements, typically about one atom in 10^7. This change is known as *doping*. Silicon and germanium have four outer electrons, these being responsible for bonding. If an element having five outer electrons, i.e., a group V element from the

Table 7.2 Numbers of conduction electrons and holes in intrinsic semiconductors at 300 K

Material	Numbers per m^3, i.e. n_i	
	Electrons n	*Holes* p
Silicon	1.4×10^{16}	1.4×10^{16}
Germanium	2.4×10^{19}	2.4×10^{19}
Gallium arsenide	1.7×10^{12}	1.7×10^{12}

Periodic table such as arsenic, antimony or phosphorus, is substituted for some silicon atoms then four of the five electrons settle into the covalent bonds with neighbouring silicon atoms, leaving a fifth electron only fairly loosely attached (Fig. 7.9(*a*)). This electron is thus fairly easily made available for conduction and since it is donating electrons for conduction the group V element is called a *donor*. The effect of this on the energy bands for the silicon is to introduce the donor electron in energy levels in the energy gap between the valence and conduction bands just below the conduction band (Fig. 7.9(*b*)). The energy gap between the donor energy levels and the conduction band is about 0.01 to 0.05 eV . Thus at room temperature virtually all the donor electrons have moved into the conduction band. There are thus more electrons in the conduction band than holes in the valence band and hence most of the electrical conduction under the action of an electric field is by electrons. For this reason this form of doped semiconductor is known as *n-type*, the n being because the conduction is by predominantly negative charge carriers, i.e., electrons.

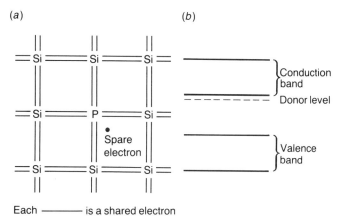

Fig. 7.9 Silicon doped with phosphorus, (*a*) structure, (*b*) energy bands

If an element having three outer electrons, i.e. a group III element from the Periodic table such as boron, aluminium, indium, or gallium, is substituted for some silicon atoms then all three of the electrons settle into covalent bonds with neighbouring silicon atoms. But there is a deficiency of one electron and so there is one bond with a silicon atom which is incomplete (Fig. 7.10(*a*)). A hole has been introduced. For this reason the group III elements are referred to as *acceptors*. The effect of this on the energy bands is to introduce the acceptor holes as vacant energy levels in the energy gap between the valence and conduction bands just above the valence band (Fig. 7.10(*b*)). The energy gap between the valence band and the acceptor energy levels is about 0.005 to

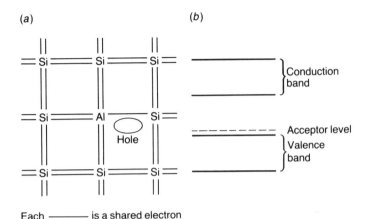

Fig. 7.10 Silicon doped with aluminium, (a) structure, (b) energy bands

0.05 eV. There are thus now more holes than conduction electrons and so most of the electrical conduction under the action of an electric field is by holes. Since the holes can be considered to be effectively positive charge carriers the material is referred to as a *p-type* semiconductor.

It is not only semiconducting elements which can be doped to become n-type or p-type semiconductors, so also can compounds. Thus, for example, gallium arsenide is a semiconducting compound between a group III element, gallium, and a group V element, arsenic. An n-type semiconductor requires a substitution by atoms which donate electrons. This can be done by replacing gallium atoms by atoms from a group IV element such as silicon, germanium or tin. Alternatively arsenic atoms can be replaced by atoms from a group VI element such as sulphur or selenium. A p-type semiconductor requires a substitution by atoms which are acceptors. This can be done by replacing gallium atoms by atoms from a group II element such as zinc, cadmium or magnesium. Alternatively arsenic atoms can be replaced by atoms from a group IV element such as germanium or silicon.

Such doped semiconductors are called *extrinsic*. With extrinsic semiconductors there will be a majority charge carrier and a minority charge carrier. With n-type material the majority charge carrier is electrons in the conduction band, with p-type it is holes in the valence band. When the material is doped the number of charge carriers introduced is virtually equal to the number of donor or acceptor atoms introduced, since each introduces one virtually free electron or hole. Typically the doping replaces about one in 10^7 atoms. Since there are about 10^{28} atoms per cubic metre, about 10^{21} virtually free electrons or holes are introduced per cubic metre. In intrinsic silicon the number of holes and the number of conduction electrons is about 10^{16} per cubic metre. Thus the doping introduces considerably more charge carriers and swamps the intrinsic population. Since some electrons and holes inevitably meet up and cancel each other out, the number of majority charge carriers is typically about 10^{21} per cubic metre and the number of minority charge carriers about 10^{10} per cubic metre.

The relationship between the number of conduction electrons n and the number of holes p is given by equation [1], i.e.,

$$np = n_i^2$$

Thus an increase in n means a decrease in p, and vice versa.

Example 2

Which of the following will be n-type and which p-type semiconductors?
(*a*) Silicon doped with gallium.
(*b*) Silicon doped with arsenic.
(*c*) Germanium doped with phosphorus.
(*d*) Germanium doped with indium.
(*e*) Gallium arsenide with sulphur replacing some arsenic atoms.

Answer

(*a*) Gallium is a group III element and so has three valence electrons, hence it is an acceptor and so the material is p type.
(*b*) Arsenic is a group V element and so has five valence electrons, hence it is a donor and so the material is n type.
(*c*) Phosphorus is a group V element and so has five valence electrons, hence it is a donor and so the material is n type.
(*d*) Indium is a group III element and so has three valence electrons, hence it is an acceptor and so the material is p type.
(*e*) Arsenic, a group V element with five valence electrons, is replaced by sulphur, a group VI element with six valence electrons, hence it is a donor and so the material is n type.

Example 3

Calculate the conduction electron and hole densities at 300 K in silicon doped with 10^{19} donors per cubic metre if the intrinsic carrier concentration is 1.4×10^{16} per cubic metre.

Answer

Using equation [1]

$$np = n_i^2$$

Since the number of donor electrons will be much greater than the number of conduction electrons it is a reasonable approximation to take n to be 10^{19} per cubic metre. Thus

$$p = \frac{n_i^2}{n} = \frac{(1.4 \times 10^{16})^2}{10^{19}} = 2.0 \times 10^{13} \text{ /m}^3$$

Thus the number of conduction electrons is about 10^{19} /m^3 and the number of holes 2.0×10^{13} /m^3. Conduction electrons are about half a million times more numerous than holes.

Currents in metals

Electrical conduction for metals, i.e., good conductors, involves the movement of the valence, or free, electrons. A simple picture of a metal is of an array of metal ions, which are in continual oscillation about their fixed positions because of thermal energy, with the free electrons continually having collisions with them and as a consequence moving about in a random manner within the metal. Between collisions the electrons move with a constant velocity. At a collision the electron exchanges energy with the ion and changes its velocity. These velocities are completely random in direction.

When an electric field is applied to the metal, the electrons are given a velocity component in the opposite direction to the field (opposite because the field direction is defined in terms of the direction of the force acting on a positive charge). The electrons thus acquire a drift velocity in this direction, this being superimposed on the random velocity. Consider the current through the conductor shown in Fig. 7.11. If the mean value of the drift velocity is v then in a time t all the electrons in a volume vtA will have moved through a cross-section of the conductor. A is the cross-sectional area of the conductor. If there are n free electrons per unit volume, each with a charge q, then the amount of charge moved through the cross-section in time t is $vtAnq$. Since current I is the rate of movement of charge then

$$I = vAnq \qquad [2]$$

This equation is often written in terms of the current density J, this being the current per unit cross-sectional area. Hence

$$J = \frac{I}{A} = vnq \qquad [3]$$

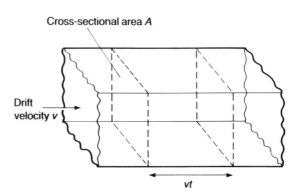

Fig. 7.11 Current through a conductor

Note that the velocity in the above equations is the drift velocity. Superimposed on that, for an electron, will be a random velocity due to thermal energy. The random kinetic energy of an electron is approximately kT, where k is Boltzmann's constant and T the temperature on the kelvin scale. Thus the random velocity v_r is given by

$$\tfrac{1}{2}mv_r^2 \approx kT$$

At 300 K, with k being 1.38×10^{-23} J/K, then v_r is about 10^5 m/s.

Example 4

What is the drift velocity of the electrons in a copper wire when a current of density 1.0×10^4 A/m^2 is flowing. Copper has 8.5×10^{28} electrons per cubic metre and the charge on an electron is 1.6×10^{-19} C.

Answer

Equation [3] gives

current density $J = vnq$

Hence

$$v = \frac{J}{nq} = \frac{1.0 \times 10^4}{8.5 \times 10^{28} \times 1.6 \times 10^{-19}} = 7.4 \times 10^{-7} \text{ m/s}$$

Conductivity of metals

The resistivity ϱ of a conductor is given by

$$R = \frac{\varrho L}{A}$$

where L is the length of the conductor, A its cross-sectional area and R its resistance. Since $V = IR$, then

$$V = \frac{I\varrho L}{A}$$

But the electric field E in the material as a result of the potential difference V across a length L is the potential gradient V/L. Hence

$$E = \frac{I\varrho}{A}$$

Instead of using the resistivity ϱ the equation can be written in terms of the conductivity σ. Since $\sigma = 1/\varrho$, then the equation becomes

$$E = \frac{I}{A\sigma} \qquad [4]$$

The current per unit area is known as the current density J, thus

$$J = E\sigma \tag{5}$$

Equation [2], $I = vAnq$, can be combined with equation [4] to give another relationship. Substituting for I in equation [4] using equation [2]

$$E = \frac{vnq}{\sigma}$$

or

$$\sigma = \frac{vnq}{E}$$

The term v/E, i.e., the drift velocity per unit electric field, is called the *mobility* μ. Mobility is a measure of the ease of carrier motion within a material. Thus

$$\mu = \frac{\text{drift velocity}}{\text{electric field}} \tag{6}$$

and so

$$\sigma = nq\mu \tag{7}$$

As a consequence of the electric field the electrons are colliding with metal ions, then accelerating under the action of the electric field until they again suffer a collision. Then they accelerate again until a further collision occurs. The electron drift velocity thus varies in a sawtooth manner (Fig. 7.12). The average time between collisions τ is called the *relaxation time*. The drift velocity v is the average velocity, thus

$$v = a\tau$$

where a is the acceleration. Using $F = ma$, then the force F acting on the electron as a result of the electric field is given by

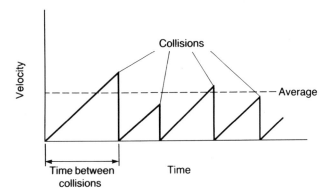

Fig. 7.12 Variation of velocity with time for an electron

$$v = \frac{F\tau}{m}$$

But the electric field strength E is F/q, hence

$$v = \frac{Eq\tau}{m}$$

and thus using equation [6], i.e., mobility $\mu = v/E$, then

$$\mu = \frac{q\tau}{m} \qquad \cdot [8]$$

The mobility is thus proportional to the relaxation time. The conductivity can thus be expressed, by the use of equations [7] and [8] as

$$\sigma = \frac{nq^2\tau}{m} \qquad [9]$$

The conductivity is thus proportional to the relaxation time.

In the above discussion the drift velocity v has been taken as related to the relaxation time by

$$v = a\tau$$

where a is the acceleration produced by the electric field. This assumes that the electron motion is acceleration followed by collision followed immediately by acceleration again (as in Fig. 7.12). But suppose the motion is acceleration followed by collision and then a delay before acceleration again (Fig. 7.13). Such delays will reduce the overall drift velocity. Such delays can occur because the electron becomes temporarily bonded to an atom for a while before breaking free to continue with conduction. Copper has a very high conductivity and its conduction electrons rarely become trapped in such bonds. Titanium, however, has many such traps for its electrons and so has a higher resistivity than copper.

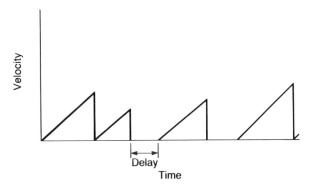

Fig. 7.13 Delays reducing drift velocity

Example 5

If the resistivity of copper is 1.6×10^{-8} Ω m what is the relaxation time for the electrons?

Charge on the electron = 1.6×10^{-19} C

Mass of the electron = 9.1×10^{-31} kg

Number of conduction electrons = 8.5×10^{28} per cubic metre

Answer

Using equation [9]

$$\sigma = \frac{nq^2\tau}{m}$$

and since the resistivity is $(1/\sigma)$ then

$$\tau = \frac{9.1 \times 10^{-3}}{1.6 \times 10^{-8} \times 8.5 \times 10^{28} \times (1.6 \times 10^{-19})^2} = 2.6 \times 10^{-14} \text{ s}$$

Example 6

What is the mobility of electrons in copper which have resistivity 1.6×10^{-8} Ω m if it has 8.5×10^{28} electrons per cubic metre?

Charge on the electron = 1.6×10^{-19} C

Answer

Using equation [7]

$$\sigma = nq\mu$$

with the conductivity σ being the reciprocal of the resistivity, then

$$\text{mobility } \mu = \frac{1}{1.6 \times 10^{-8} \times 8.5 \times 10^{28} \times 1.6 \times 10^{-19}}$$

$$= 4.6 \times 10^{-3} \text{ m}^2 \text{ V}^{-1} \text{ s}^{-1}$$

Example 7

What is the electric field in copper of resistivity 1.6×10^{-8} Ω m when the current density is 5.0×10^6 A/m²?

Answer

Using equation [5]

$$J = E\sigma$$

and, since the conductivity is the reciprocal of the resistivity,

$$E = 5.0 \times 10^6 \times 1.6 \times 10^{-8}$$

$$= 8.0 \times 10^{-2} \text{ V/m}$$

Effect of temperature on metal conductivity

Figure 7.14 shows how the resistivity varies with temperature for some metals. All would appear to indicate zero resistivity at 0 K, the resistivity then increasing as the temperature is increased.

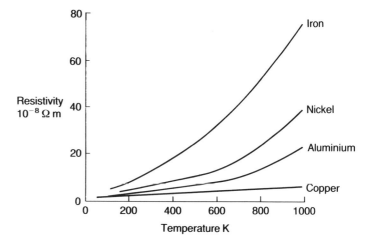

Fig. 7.14 Resistivity variation with temperature for some metals

As indicated by equation [9] the conductivity of a metal is proportional to the relaxation time, i.e., the average time between collisions of an electron with the lattice of metal ions in which there is a transfer of energy. When the temperature of a metal is increased the atoms vibrate more about their mean positions and so impede the movement of the electrons more. Thus the time between collisions will be reduced. A reduction in the relaxation time means a reduction in the conductivity (an increase in resistivity) and thus an increase in temperature means a reduction in conductivity.

The above is a very simple picture of the effects of temperature on the conductivity of metals. The conduction electrons in their movement through the lattice of metal atoms collide with very few of them. For instance, in copper at about room temperature the random velocity of the electrons is about 10^5 m/s (see earlier in this chapter) while the relaxation time is about 10^{-14} s (see example 5). Thus the average distance travelled by an electron between collisions is $10^5 \times 10^{-14}$ m or about 10^{-9} m. Atoms are about 10^{-10} m in diameter and thus the electron moves on average about 10 atomic diameters through the densely packed lattice without suffering a collision. This argument has considered electrons as particles bouncing about between atoms. A better model is however to consider the electron as a wave which passes through much of the lattice without interaction. Indeed in a perfectly orderly array we can think of the wave passing through the lattice without any interaction. The interaction that does occur is with the wave packets of vibrational energy emanating from the atoms as a result of thermal energy. We can think of the situation being rather like a crowd of closely packed people. If one person starts rocking back-and-forth then all the neighbouring people become affected and they in turn cause yet more people to become affected. Thus a wave motion spreads out from the person. With the metal the spread out of the vibrational energy thus disrupts the orderly array of the atoms in the lattice and consequently leads to interaction with the electron wave. These vibrational wave packets are called *phonons*, the total number of such phonons, i.e., the total vibrational energy, being proportional to the temperature on the kelvin scale. Thus, in the absence of phonons, i.e., at a temperature of absolute zero, the electron wave should pass through the lattice as though it were not there and so there would be no resistance. An increase in temperature means an increase in the number of phonons and hence an increased number of 'collisions'. Thus increasing the temperature increases the resistance.

Superconductivity

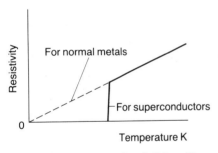

Fig. 7.15 Variation of resistivity with temperature for normal metals and superconductors

The resistivity of metals decreases as the temperature decreases and by extrapolation could be expected to become zero at a temperature of absolute zero, i.e., 0 K. However, for some metals, the resistivity suddenly drops to zero at a temperature above absolute zero (Fig. 7.15). Such materials are said to be *superconductors*. The temperature at which this abrupt change to zero resistivity occurs is called the *critical temperature* T_c. The superconductivity can be destroyed if a superconductor, below the critical temperature, is in a magnetic field which is above some critical value or the current density in the material is above some critical value (we can think of this as being an internally generated magnetic field).

Until 1986 the critical temperatures of superconductors had all tended to be very low, for many less than about 10 K with the highest being 23 K, and the superconductivity was destroyed by relatively low magnetic fields. However, in 1986 high-temperature superconductivity was announced with a compound which had a critical temperature of 30 K and required a relatively high magnetic field to destroy it. Since then more superconducting compounds have been discovered with yet higher critical temperatures, some with critical temperatures higher than 100 K.

The current in normal conductors is carried by 'normal' electrons which individually pass through the lattice. In a collision with an ion there is a transfer of energy. With superconductors, however, some electrons travel in pairs. We can picture the situation to be rather like one electron interacting with a metal ion so that the electron loses a phonon, i.e., a packet of energy. The other electron happens to be in just the right place and moving at the right speed to 'catch' this phonon. Thus the net effect on the pair of electrons is of no change. The pair can thus travel through the metal continually interchanging phonons and suffer no net loss in energy. A magnetic field can break up this pair arrangement by trying to bend the tracks of the electrons.

Electrical conductivity with semiconductors

Semiconductors can have both electrons and holes as charge carriers for electrical conduction. Thus the equation used for the conductivity of metals where there were only electrons has to be modified (eqn [7] conductivity $\sigma = nq\mu$ with n being the number of conduction electrons per cubic metre, q the charge carried by an electron and μ the mobility). It becomes

$$\sigma = nq\mu_n + pq\mu_p \qquad [10]$$

where n is the number of electrons with mobility μ_n and p the number of holes with mobility μ_p.

In intrinsic materials the number of electrons n equals the number of holes p, with both thus equalling the intrinsic carrier concentration n_i. Thus equation [10] becomes

$$\sigma = n_i q(\mu_n + \mu_p) \qquad [11]$$

Table 7.3 gives the values of mobilities in intrinsic semiconductor materials at 300 K.

Table 7.3 Mobilities of electrons and holes at 300 K

Material	Mobilities m^2 V^{-1} s^{-1}	
	electrons	holes
Germanium	0.39	0.19
Silicon	0.15	0.048
Gallium arsenide	0.85	0.048
Gallium phosphide	0.045	0.0020
Indium antimonide	8.00	0.020

In extrinsic materials the effect of the doping is to make either electrons or holes the major carrier and thus one of the terms in equation [10] can generally be neglected. Thus for n-type material when the carriers are overwhelmingly electrons

$$\sigma = nq\mu_n \qquad [12]$$

and for p-type material when the charge carriers are overwhelmingly holes

$$\sigma = pq\mu_p \qquad [13]$$

The number of majority charge carriers is proportional to the dopant concentration. Thus we might expect the conductivity to be proportional to the dopant concentration. This is the case at low dopant concentrations, less than 0.001%, because the dopant concentration is not large enough significantly to affect the mobilities. At higher dopant concentrations the mobility is significantly affected by the presence of the dopant ions because of the distortion they produce in the lattice structure.

Example 8

What is the conductivity of intrinsic silicon at 300 K if the intrinsic carrier concentration is 1.4×10^{16} /m^3 and the mobilities are as given in Table 7.3? Take the charge carried by the electrons and holes to be 1.6×10^{-19} C.

Answer

Using equation [11]

$$\sigma = n_i q(\mu_n + \mu_p)$$
$$= 1.4 \times 10^{16} \times 1.6 \times 10^{-19}(0.15 + 0.048)$$
$$= 4.4 \times 10^{-4} \Omega^{-1} \, m^{-1}$$

Example 9

What is the conductivity of silicon doped with boron at 300 K if the number of boron atoms is 10^{21} /m^3 and the mobility of the holes is 0.048 m^2 V^{-1} s^{-1}? Take the charge carried by a hole to be 1.6×10^{-19} C.

Answer

Using equation [13]

$$\sigma = pq\mu_p$$

with the number of holes being taken to be equal to the number of boron atoms, then

$$\sigma = 10^{21} \times 1.6 \times 10^{-19} \times 0.048 = 77 \; \Omega^{-1} \, m^{-1}$$

Compensated semiconductors

A semiconductor which has been doped by both donor and acceptor impurities is said to be *compensated*, being fully compensated if there are equal concentrations of donor and acceptor dopants. With such a semiconductor the holes supplied by the acceptors are promptly occupied by the free electrons supplied by the donors. In the fully compensated semiconductor the material will thus only have the same number of charge carriers as in the intrinsic semiconductor. The mobilities of the electrons and holes are however lower than in the intrinsic material because the presence of the donor ions in the lattice distorts it.

Fermi level

The energy band picture of a solid is of a valence band containing the outermost atomic electrons and a conduction band. In the case of a semiconductor, at a temperature of 0 K the valence band is full and there is a small energy gap to the empty conduction band. The probability that a particular level of energy E will be occupied by an electron is given by

$$p(E) = \frac{1}{1 + \exp[(E - E_F)/kT]} \tag{14}$$

E_F is called the Fermi energy or more commonly the *Fermi level*, k is Boltzmann's constant and T the temperature on the kelvin scale. The term $p(E)$ means the probability of energy E.

As a brief digression, a coin has two ways of landing – heads uppermost or tails uppermost. Thus when a coin is thrown onto a table it can land in one of two ways. Thus there is a 1 in 2 probability that it will land heads uppermost, i.e., the probability is ½. The probability of it landing with either heads or tails uppermost is 2 in 2, i.e., a probability of 1. A probability of 1 is a certainty.

At $T = 0$ K, the probability of an electron being in a particular level of energy E which is below E_F is given by

$$p(E) = \frac{1}{1 + \exp(-\text{ infinity})}$$

Since the value of $\exp(-\text{ infinity})$ is 0 then the probability is 1. This means it is a certainty that the levels below the Fermi level will be occupied.

The probability of an electron, at 0 K, being in a level E which is greater than E_F is given by

$$p(E) = \frac{1}{1 + \exp(+\text{ infinity})}$$

Since the value of $\exp(+\text{ infinity})$ is infinity then the probability is 0. A probability of zero means that there is no chance at all of any energy level with energy greater than the Fermi level being occupied. Thus at 0 K all the energy levels below the Fermi level are occupied and all above it are empty, as illustrated by Fig. 7.16.

At temperatures other than 0 K, when $E = E_F$ then

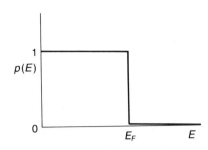

Fig. 7.16 The probability of occupancy of energy levels at 0 K

$$p(E_F) = \frac{1}{1 + \exp(0)}$$

Since the value of $\exp(0)$ is 1 then the probability is ½. Figure 7.17 shows how the probability of occupancy of energy levels varies with temperature, all however have the probability of ½ at the Fermi level. At each temperature the probability graph is symmetrical about the Fermi level. This is because the electrons moved from levels below E_F are all moved to levels above E_F.

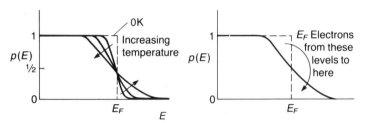

Fig. 7.17 The probability of occupancy of energy levels at temperatures other than 0 K

In the case of a metal the Fermi level is the top of the occupied energy levels at 0 K and so at higher temperatures there are as many electrons moved above the level as gaps in the energy levels below it.

Work function

In Chapter 6 thermionic and photoelectric emission of electrons from solids was discussed and the term work function ϕ introduced, the *work function* being the minimum energy required at 0 K to cause the emission of electrons from the surface of a solid. An alternative, and equivalent, definition of the work function is the energy that has to be given to electrons at the Fermi level for them to escape from the surface.

Electron and hole density

If we want to count the number of electrons in a particular band of energies, e.g., the conduction band, then we have to look at the probabilities for each level. Thus if for energy level E_1 in the band the probability is 1 it is certain we will find one electron there. If for energy level E_2 the probability is ½ then there is a 1 in 2 chance of finding an electron there and so over a period of time the average number of electrons there is ½. Thus to find the total number of electrons in an energy band we have to sum all the values of $p(E)$ for values of E possible within the band, i.e., for each of the energy levels within the band. This is

$$\frac{1}{1 + \exp[(E_1 - E_F)/kT]} + \frac{1}{1 + \exp[(E_2 - E_F)/kT]} + \text{etc.}$$

If we take the conduction band as having energies extending from E_c, the bottom of the band, to infinity then such a summation yields the result that the number of electrons n per cubic metre in the conduction band is given by

$$n = N_c \exp[-(E_c - E_F)/kT] \qquad [15]$$

where N_C is the effective number of energy levels in the conduction band per cubic metre.

The probability of throwing a six-sided die and getting a six uppermost is 1 in 6, i.e., 1/6. The probability of not getting a six is 5 in 6, i.e. $(1 - 1/6)$. Thus the probability of finding a hole at an energy level in the valence band is 1 minus the probability of there being an electron there. Since

$$1 - \frac{1}{1 + \exp[(E_1 - E_F)/kT]} = \frac{1}{1 + \exp[(E_F - E_1)/kT]}$$

the summation of all the terms within the valence band leads to the number of holes p per cubic metre being

$$p = N_V \exp[-(E_F - E_V)/kT] \qquad [16]$$

where E_V is the energy at the top of the valence band and N_V the effective number of energy levels in the valence band.

The Fermi level for intrinsic semiconductors

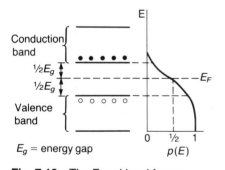

E_g = energy gap

Fig. 7.18 The Fermi level for an intrinsic semiconductor

For an intrinsic semiconductor there are equal numbers of electrons in the conduction band and holes in the valence band. Because the probability, and hence number of electrons above the level and holes below it, is symmetrical about the Fermi level then the Fermi level must fall in about the middle of the energy gap between the valence and conduction bands (Fig. 7.18).

Because the number of conduction electrons per cubic metre n equals the number of holes per cubic metre p then equations [15] and [16] give

$$n = p$$

$$N_C \exp[-(E_C - E_F)/kT] = N_V \exp[-(E_F - E_V)/kT]$$

$$\frac{N_C}{N_V} = \frac{\exp[-(E_F - E_V)/kT]}{\exp[-(E_C - E_F)/kT]}$$

$$= \exp[(-E_F + E_V + E_C - E_F)/kT]$$

$$\ln(N_C/N_V) = (-2E_F + E_V + E_C)/kT$$

Hence

$$E_F = \tfrac{1}{2}(E_C + E_V) + \tfrac{1}{2}kT \ln(N_C/N_V) \qquad [17]$$

The $\tfrac{1}{2}kT \ln(N_C/N_V)$ term is very small and thus the Fermi level for an intrinsic semiconductor is about half-way between the valence and conduction bands.

The Fermi level for extrinsic semiconductors

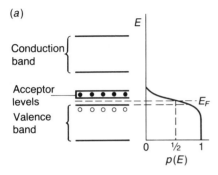

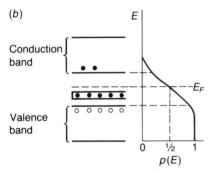

Fig. 7.19 The Fermi level of a p-type semiconductor at (a) low and (b) higher temperatures

In an extrinsic semiconductor there are not equal numbers of electrons and holes. For a p-type semiconductor acceptor energy levels are introduced in the gap between the valence and conduction bands. At low temperatures the conduction is by holes in the valence band resulting from electrons having jumped from the valence band to the acceptor energy levels, few making it to the conduction band (Fig. 7.19(a)). Thus for this situation the Fermi level must lie between the valence band and the acceptor levels. At higher temperatures some electrons will make it to the conduction band. As a consequence the Fermi level will move upwards towards the middle of the energy gap (Fig. 7.19(b)). The Fermi level thus lies in the bottom half of the energy gap.

For an n-type semiconductor donor energy levels are introduced in the gap between the valence and conduction bands. At low temperatures the conduction is by electrons which have jumped from the donor levels to the conduction band, few making it from the valence band (Fig. 7.20(a)). Thus for this situation the Fermi level must lie between the donor levels and the conduction band. At higher temperatures some of the electrons will make it from the valence band. As a consequence the Fermi level will move downwards towards the middle of the energy gap (Fig. 7.20(b)). The Fermi level thus lies in the upper half of the energy gap.

The position of the Fermi level in n- and p-type semiconductors is also affected by the amount of doping. With no doping the Fermi level would be about half-way between the valence and conduction bands. Introducing donors moves the Fermi level towards the conduction band, the greater the extent of the doping the more the level moves. If the doping density is made high enough there are so many doping atoms in the lattice that the Fermi level can get so close to the conduction level, or even reach it, that the semiconductor behaves like a metal as a consequence of the large number of electrons donated by the donor atoms. The semiconductor is said to have degenerated to a metal and is called *degenerate*. Introducing acceptors moves the Fermi level towards the valence band, the greater the extent of the doping the more the level moves. With high enough doping the Fermi level can move close enough to, or even reach, the valence level and as a consequence the semiconductor becomes degenerate. For silicon at room temperature the maximum non-degenerate doping concentrations are about 1×10^{24} per cubic metre for donors and 9×10^{23} per cubic metre for acceptors.

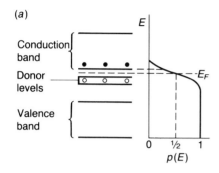

(a)

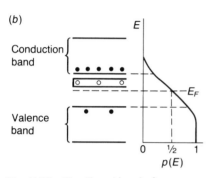

(b)

Fig. 7.20 The Fermi level of an n-type semiconductor at (a) low and (b) higher temperatures

For an intrinsic semiconductor the conduction electron density n equals the hole concentration density p, the concentration being known as the intrinsic concentration n_i.

$$n = p = n_i$$

Thus equation [15] gives

$$n_i = n = N_C \exp[-(E_C - E_{Fi})/kT]$$

where E_{Fi} is the position of the intrinsic Fermi level. For an extrinsic semiconductor equation [15] gives

$$n = N_C \exp[-(E_C - E_F)/kT]$$

Thus dividing these two equations gives

$$\frac{n}{n_i} = \frac{N_C \exp[-(E_C - E_F)/kT]}{N_C \exp[-(E_C - E_{Fi})/kT]}$$

$$\frac{n}{n_i} = \exp[(- E_C + E_F + E_C - E_{Fi})/kT]$$

$$n = n_i \exp[(E_F - E_{Fi})/kT]$$

Hence

$$E_F - E_{Fi} = kT \ln(n/n_i) \qquad [18]$$

Thus the greater the ratio n/n_i the further out the Fermi level is from the intrinsic Fermi level.

A similar expression can be derived for acceptor doped semiconductors.

$$E_F = E_{Fi} = kT \ln(p/n_i) \qquad [19]$$

Example 10

What is the position, relative to the intrinsic Fermi level, of the Fermi level in silicon at 300 K doped with 10^{21} donors per cubic metre if the intrinsic concentration is 1.4×10^{16} /m³?
Boltzmann's constant $k = 1.38 \times 10^{-23}$ J/K

Answer

Using equation [18] and assuming that each donor atom has released one electron,

$$E_F - E_{Fi} = kT \ln(n/n_i)$$
$$= 1.38 \times 10^{-23} \times 300 \ln(10^{21} \times 1.4 \times 10^{16})$$
$$= 4.6 \times 10^{-20} \text{ J} = 0.29 \text{ eV}$$

The Fermi level is 0.29 eV above the intrinsic Fermi level, i.e., effectively 0.29 eV above the centre of the energy gap between the valence and conduction bands.

Example 11

A semiconductor is said to be degenerate when, in the case of a donor doped semiconductor, the Fermi level is located at an energy of 3 kT or less from the conduction band. Calculate the minimum doping density for silicon at 300 K to become degenerate if n_i is 1.4×10^{16} /m³ and the energy gap between the valence and conduction bands is 1.1 eV?
Boltzmann's constant $k = 1.38 \times 10^{-23}$ J/K

Answer

Using equation [18]

$$E_F - E_{Fi} = kT \ln(n/n_i)$$

But, to a reasonable approximation, $E_{Fi} = (E_C + E_V)/2$, thus

$$E_F - (E_C + E_V)/2 = kT \ln(n/n_i)$$

For $E_F = E_C - 3kT$, then

$$E_C - 3kT - (E_C + E_V)/2 = kT \ln(n/n_i)$$

$$(E_C - E_V)/2 - 3kT = kT \ln(n/n_i)$$

But the energy gap is $(E_C - E_V)$, hence

$$(1.1 \times 1.6 \times 10^{-19})/2 - 3 \times 1.38 \times 10^{-23} \times 300$$

$$= 1.38 \times 10^{-23} \times 300 \ln(n/1.4 \times 10^{16})$$

Thus n is 1.2×10^{24} /m³.

Effect of temperature on intrinsic semiconductor conductivity

For an intrinsic semiconductor the number of holes p in the valence band equals the number of electrons n in the conduction band,

$$n = p = n_i$$

where n_i is the intrinsic carrier concentration. But, equation [1] gives

$$np = n_i^2$$

thus, using equations [15] and [16],

$$n_i^2 = N_V N_C \exp[-(E_F - E_V)/kT]\exp[-(E_C - E_F)/kT]$$

$$n_i^2 = N_V N_C \exp[-(E_C - E_V)/kT]$$

$$n_i^2 = N_V N_C \exp[- E_g/kT]$$

where E_g is the energy gap between the valence and conduction bands. Hence n_i is given by the square root of the above expression, i.e.,

$$n_i = A \exp[- E_g/2kT]$$

where A is a constant. The electrical conductivity σ of the semiconductor is given by equation [11] as

$$\sigma = n_i q(\mu_n + \mu_p)$$

An increase in temperature results in a movement of electrons from the valence to the conduction band and hence an increase in n_i. Though there is a decrease in mobility with an increase in temperature the increase in n_i produces a much greater change in conductivity and is largely responsible for the way the conductivity changes with temperature. Hence for a particular material the conductivity is reasonably proportional to n_i, the other terms in comparison being little affected by a change in temperature. But n_i is proportional to $\exp[-E_g/2kT]$, thus

$$\sigma = B \exp[- E_g/2kT] \qquad [20]$$

where B is a constant.

Experimental measurements of the electrical conductivity of an intrinsic material at different temperatures enables the size of the energy gap between the valence and conduction bands to be determined. Equation [20] can be written as

$$\ln \sigma = \ln B - [E_g/2kT] \qquad [21]$$

Thus a graph of $\ln \sigma$ plotted against $1/T$ gives a straight-line graph (Fig. 7.21) with a slope of $-E_g/2k$. Hence E_g can be obtained from the graph.

Example 12

Figure 7.22 shows how the logarithm of the resistance of a sample of germanium varies with the reciprocal of the absolute temperature.

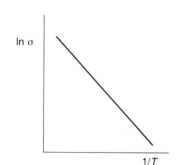

Fig. 7.21 Graph of conductivity with temperature for an intrinsic semiconductor

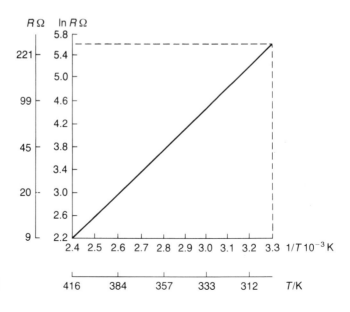

Fig. 7.22 Example 12

What value does the graph indicate for the energy gap in germanium? Boltzmann's constant $= 1.38 \times 10^{-23}$ J/K

Answer

Equation [21] gives the equation for the graph in terms of the conductivity,

$$\ln \sigma = \ln B - [E_g/2kT]$$

Since the conductivity σ is L/RA, where L is the length of the sample, A its cross-sectional area and R its resistance,

$$\ln (L/RA) = \ln B - [E_g/2kT]$$

$$\ln (L/A) - \ln R = \ln B - [E_g/2kT]$$

$$\ln R = E_g/2kT - \ln B + \ln (L/A)$$

$$\ln R = E_g/2kT - \text{a constant}$$

The slope of the graph is thus $E_g/2k$. Hence

$$E_g = 2 \times 1.38 \times 10^{-23} \times (3.5/0.9 \times 10^{-3})$$

$$= 1.07 \times 10^{-19} \text{ J} = 0.67 \text{ eV}$$

Effect of temperature on extrinsic semiconductor conductivity

At low temperatures, below about 100 K, the temperatures are not high enough to move many electrons from the valence band to the conduction band. Thus for an n-type semiconductor most of the conduction is due to electrons moving from the donor levels into the conduction band (Fig. 7.23(a)), the number of conduction electrons being less than the number of donor atoms. The effect of increasing the temperature is thus to increase the number of electrons in the conduction band as a result of more donor atoms losing electrons. At higher temperatures all the donor atoms have lost their electrons and though more electrons are able to jump from the valence band to the conduction band the conduction is mainly due to the electrons from the donor atoms (Fig. 7.23(b)). The number of electrons for conduction is thus effectively equal to the number of donor atoms and since this is constant the number

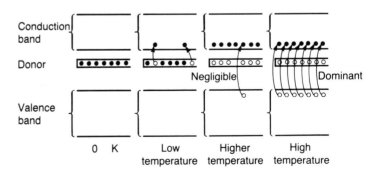

Fig. 7.23 The effect of temperature on conduction in an n-type semiconductor

of conduction electrons is essentially constant. In this region the increase in temperature decreases the conductivity because of the decrease in mobility of the electrons. The conductivity is given by equation [12], i.e.,

$$\sigma = nq\mu_n$$

with n constant and μ_n decreasing. At yet higher temperatures the numbers of electrons moving from the valence band to the conduction band increases and can become greater than those from the donor atoms and can swamp the effect of the donor atoms (Fig. 7.23). The material thus becomes effectively an intrinsic semiconductor, the conductivity then being a result of almost equal numbers of electrons in the conduction band and holes in the valence band. The conductivity is then effectively given by equation [10]

$$\sigma = nq\mu_n + pq\mu_p$$

with n approaching n_i as the temperature increases. Both n and p are increasing and in comparison the decrease in mobilities becomes insignificant. Hence the conductivity increases. Figure 7.24 shows how the conductivity tends to vary with temperature for an extrinsic semiconductor.

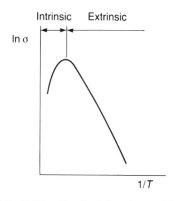

Fig. 7.24 Conductivity of an extrinsic semiconductor

Hall effect

A current flowing in a conductor can be deflected by a magnetic field in just the same way as a beam of electrons moving in a vacuum can be deflected by a magnetic field (see Ch. 6). This effect was first shown by E.H. Hall in 1879 and is now known as the *Hall effect*. The effect occurs in both metals and semiconductors and can be used to determine the current carrier density, i.e., the number per cubic metre, and mobility.

Consider a plate of metal or semiconductor of thickness t, length L and width w (Fig. 7.25) with a uniform magnetic field of flux density B normal to the plane of the plate. If a steady current I is carried through the plate by charge carriers, each having a charge q (this would be a negative charge for electrons and positive for holes), then the force F exerted on each charge carrier will be (see Ch. 7 eqn [9])

$$F = Bqv$$

where v is the velocity of the charge carriers in the direction of the current. This force will cause the charge carriers to be deflected so that they crowd to one edge of the plate. The side they go to depends on whether the charge carriers are electrons or holes (see Fig. 7.25). The result is that one edge of the plate becomes positively charged and the opposite edge negatively charged. This charge separation will produce an electric field E which will exert forces on the charge carriers in

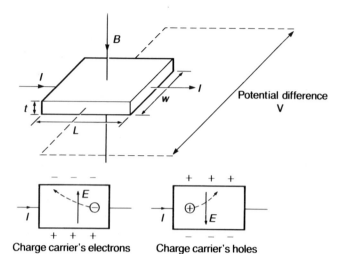

Fig. 7.25 The Hall effect

Charge carrier's electrons   Charge carrier's holes

the opposite direction to that produced by the magnetic field. The force on a charge carrier due to the electric field is (see Ch. 7)

$$F = qE$$

The charge separation grows until the force due to the electric field becomes equal to the force produced by the magnetic field. Then

$$qE = Bqv$$

and so

$$E = Bv$$

The velocity v is the drift velocity of the electrons in the current and is thus given by equation [2] as

$$I = vAnq$$

where A, the cross-sectional area, is wt and n the number of charge carriers per cubic metre. Thus

$$E = \frac{BI}{wtnq} \qquad [22]$$

This equation is generally written as

$$E = R_H \frac{BI}{wt} \qquad [23]$$

where R_H is the *Hall coefficient*, being $1/nq$.

The electric field across the width of the plate means there is a potential difference V across it, with E being the potential gradient (see Ch. 2). Thus

$$E = \frac{V}{w}$$

Hence equation [22] becomes

$$V = \frac{BI}{tnq} = R_H \frac{BI}{t} \qquad [24]$$

This equation indicates one of the applications of the Hall effect. Since for a constant current through the plate the transverse potential difference V is proportional to the normal flux density B, a measurement of V can be used to obtain a measure of the flux density.

For a metal and extrinsic semiconductors where there is essentially just one charge carrier the electrical conductivity σ is given by (eqn [7] for metals)

$$\sigma = nq\mu$$

Hence, since R_H is $1/nq$

$$\sigma = \frac{\mu}{R_H} \qquad [25]$$

Thus a measurement of the conductivity and a determination of R_H by measurements involving equation [24] enables the mobility to be determined.

Example 13

A Hall probe is being used to determine the magnetic flux density at right-angles to its surface. If the probe has a thickness of 0.20 mm and contains 3.0×10^{22} charge carriers per cubic metre, what is the flux density when the Hall voltage is 100 mV with a current through the probe of 50 mA? The charge carried by each carrier is 1.6×10^{-19} C.

Answer

Using equation [24]

$$V = \frac{BI}{tnq}$$

$$B = \frac{100 \times 10^{-3} \times 0.20 \times 10^{-3} \times 3.0 \times 10^{22} \times 1.6 \times 10^{-19}}{50 \times 10^{-3}}$$

$$= 1.9 \text{ T}$$

Example 14

What is the mobility and density of doping of a semiconductor if a plate of it gives a Hall voltage of 15 mV when the current through the plate is 48 mA and the magnetic flux density at right-angles to it is 0.10 T? The plate has a thickness of 2.0 mm, a width of 4.0 mm and a length of 5.0 mm. The current through the plate is produced by a voltage of 2.0 V between its ends.

Answer

The conductivity σ of the plate is given by

$$\sigma = \frac{L}{RA}$$

with the resistance R being given by the applied voltage divided by the current. Thus, since the cross-sectional area A is 2.0×4.0 mm^2 and the length L is 5.0 mm,

$$\sigma = \frac{5.0 \times 10^{-3}}{(2.0/0.048) \times 2.0 \times 4.0 \times 10^{-6}}$$

$$= 15 \ \Omega^{-1} \ m^{-1}$$

Equation [24] gives

$$V = R_H \frac{BI}{t}$$

Hence

$$R_H = \frac{15 \times 10^{-3} \times 2.0 \times 10^{-3}}{0.10 \times 0.048}$$

$$= 6.3 \times 10^{-3} \ m^3/C$$

Thus since

$$R_H = \frac{1}{nq}$$

the number of charge carriers per cubic metre is given by

$$n = \frac{1}{6.3 \times 10^{-3} \times 1.6 \times 10^{-19}} = 9.9 \times 10^{20} \ /m^3$$

Since, equation [25],

$$\sigma = \frac{\mu}{R_H}$$

then the mobility μ is

$$\mu = 15 \times 6.3 \times 10^{-3} = 9.5 \times 10^{-2} \ m^2 \ V^{-1} \ s^{-1}$$

Problems

1 Describe the typical form of the conduction and valence energy bands for (*a*) good conductors, (*b*) semiconductors, and (*c*) insulators.

2 Silicon oxide has an energy gap between the valence and conduction bands of 8 eV, while silicon has an energy gap of about 1 eV. What will be the difference in electrical conductivity behaviour of these two materials?

3 Which of the following materials will be p-type and which n-

type semiconductors?

(*a*) Silicon doped with aluminium,

(*b*) silicon doped with phosphorus,

(*c*) germanium doped with aluminium,

(*d*) germanium doped with arsenic,

(*e*) gallium arsenide with some gallium atoms replaced by silicon,

(*f*) gallium arsenide with some arsenic atoms replaced by tellurium.

4   Calculate the electron and hole numbers per cubic metre in silicon at 300 K when doped with 10^{22} acceptor atoms per cubic metre if the intrinsic carrier concentration is 1.4×10^{16} per cubic metre.

5   What is the drift velocity of the conduction electrons in a copper wire of cross-sectional area $1.0 \times 10^{-6} \, m^2$ when there is a current of 2.0 A? The number of conduction electrons in the copper is 8.5×10^{28} per cubic metre and the charge on an electron is 1.6×10^{-19} C.

6   The resistivity of copper at 0 °C is $1.6 \times 10^{-8} \, \Omega$ m and changes to $2.2 \times 10^{-8} \, \Omega$ m at 100 °C. By what factor does the relaxation time of the conduction electrons change?

7   What is (*a*) the mobility, (*b*) the electric field strength and (*c*) the drift velocity for conduction electrons in copper with a conductivity of $5.5 \times 10^{7} \, \Omega^{-1} \, m^{-1}$, 8.5×10^{28} electrons per cubic metre and a current density of 2.0×10^{6} A/m? The charge on the electron is 1.6×10^{-19} C.

8   What is the resistivity of intrinsic germanium at 300 K if the intrinsic carrier concentration is $2.4 \times 10^{19} \, /m^3$ and the mobilities are for electrons $0.39 \, m^2 \, V^{-1} \, s^{-1}$ and holes $0.19 \, m^2 \, V^{-1} \, s^{-1}$? The charge on the electrons and holes is 1.6×10^{-19} C.

9   What is the resistivity of germanium doped with 10^{22} donor atoms per cubic metre at 300 K if the mobility of the electrons is $0.39 \, m^2 \, V^{-1} \, s^{-1}$ and the charge on the electrons is 1.6×10^{-19} C?

10   Explain what is meant by the Fermi level and give typical energy band pictures including it for intrinsic and extrinsic semiconductors, describing the effect of increased doping on its position.

11   What is the position, relative to the intrinsic Fermi level, of the Fermi level in silicon at 300 K doped with 10^{22} donors per cubic metre if the intrinsic concentration is $1.4 \times 10^{16} \, /m^3$? Boltzmann's constant $k = 1.38 \times 10^{-23}$ J/K.

12   Describe how the conductivity of (*a*) intrinsic and (*b*) extrinsic semiconductors changes with temperature and give a qualitative explanation of the changes.

13   Explain how measurements of the electrical conductivity of a semiconductor over a range of temperatures can be used to determine the energy gap between the valence and conduction bands.

14   A strip of copper carries a current of 6.0 A in a magnetic field of flux density 1.2 T at right-angles to its surface. The strip has a width of thickness of 0.10 mm and develops a Hall voltage of 5.3 μV. What is the number of charge carriers per cubic metre?

Charge on the electron $= 1.6 \times 10^{-19}$ C.

15 A current of 10 mA passes along the length of a plate of n-type germanium, thickness 0.10 mm, when there is a magnetic field of flux density 0.20 T at right-angles to it. A Hall voltage of 3.8 mV is produced. What is (a) the number of electrons per cubic metre, and (b) the Hall coefficient? Charge carried by a charge carrier $= 1.6 \times 10^{-19}$ C.

8 Junctions between materials

Introduction

This chapter is concerned with what happens when charge carriers are introduced or generated in semiconductors and consequently what happens when materials are in contact and charge carriers can move across the junction. In particular, the junctions between metals and semiconductors and those between p- and n-type semiconductors are considered. The development of electronic devices based on such junctions, such as light-emitting diodes, injection lasers, field emission transistors and bipolar transistors is briefly explored.

Excess charge carriers

The situations discussed in Chapter 7 involve only materials which are uniform throughout in composition and temperature. Thus, for example, a doped semiconductor has been assumed to be doped the same throughout. For such situations the number of charge carriers per cubic metre does not vary with position within the material or with time. In addition there has been no input of charge carriers from some external source. Thus the number of charge carriers per cubic metre has been determined only by the nature of the semiconductor material concerned, the dopant density and the temperature. The number of charge carriers under such conditions is the equilibrium carrier concentration; any carrier concentration above that is called the *excess carriers*. Excess carriers can be produced in a material as a result of it being illuminated by light or by injection through contact with some other material. Uniform illumination of the material by light can result in a uniform production of excess charge carriers throughout the material, whereas the injection of carriers through contact with some other material produces a non-uniform distribution with the excess charge carrier density being greater nearer the point of contact than elsewhere in the material.

With the equilibrium carrier concentration the situation is

not static with the same charge carriers always being present. The situation is dynamic with charge carriers being constantly created and disappearing; the thing that is constant is the overall number of charge carriers. The rate at which charge carriers are created equals the rate at which they disappear. Charge carriers disappear when a free electron meets a hole and becomes locked in the bonding between atoms instead of being free. The term *recombination* is used for this disappearance.

Recombination and lifetime

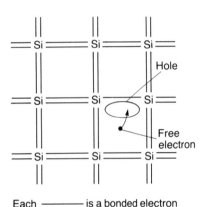

Each ———— is a bonded electron

Fig. 8.1 Recombination

Recombination occurs when free electrons fall into holes. When this happens a hole and the free electron vanish (Fig. 8.1). The electron in moving into the hole has become an electron involved in the bonding between atoms and so is no longer free. The chance of recombination by electrons falling directly from the conduction band into holes in the valence band is negligible in germanium or silicon. Instead the electrons fall between the conduction band and valence band in a number of steps via intermediate energy levels (Fig. 8.2). If the intermediate level is just below the bottom of the conduction band in an n-type semiconductor and empty, i.e., it has a hole, then it has a high chance of capturing an electron from the conduction band and is referred to as an *electron trap*. If the intermediate level is just above the top of the valence band in a p-type semiconductor and occupied by an electron it has a high chance of capturing a hole from the valence band and is thus referred to as a *hole trap*. If the intermediate level is about half-way between the valence and conduction band and occupied by an electron it has a reasonable chance of the electron moving down into the valence band, the intermediate level then becoming empty. An alternative way of putting this is that the intermediate level captures a hole from the valence band with the electron moving to the valence band and becoming lost to the conduction process. The intermediate level is then empty, i.e., has a hole, and can thus capture an electron from the conduction band. The net result is an electron in the intermediate level, as at the start, and that an electron has moved down from the conduction level and been recombined with a hole moving up from the valence band. Such an intermediate level is known as a *recombination centre*. Each time an electron drops down levels, or a hole moves upwards, energy is released. These intermediate levels result from impurity atoms and discontinuities in the orderly arrangement of the atoms in the crystal lattice.

The average time an electron or hole can exist in the free state, i.e., available for conduction and not locked up in

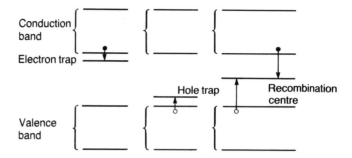

Fig. 8.2 Traps and recombination centres

bonds, is called the *lifetime* τ. Its value depends on the chance of recombination, the higher this chance the shorter the lifetime. Thus impurities and imperfections in the crystal lattice reduce the lifetime.

If n_e is the excess number of charge carriers then in a time τ all will have recombined. In half the time half will have recombined. Thus the rate of recombination is

$$\text{rate of recombination} = -\frac{n_e}{\tau} \qquad [1]$$

The minus sign is because the number calculated by this equation as recombining in some time has to be deducted from the total number starting.

Generation

Generation is the term used to describe the processes by which electrons and holes are created to add to the numbers already present in the semiconductor. One way this can be done is by illuminating the material, energy thus being supplied by photons colliding with the electrons in the material. The energy of a photon is given by $E = hf$, where h is Planck's constant and f the frequency of the light.

Generation can involve the incident energy being absorbed by an electron in the valence band and, providing the energy is large enough, causing the electron to jump to the conduction band (Fig. 8.3(a)). The result of this is that a hole has been produced in the valence band and an electron in the conduction band. An electron–hole pair has been created.

Another possible generation mechanism uses recombination centres, i.e., energy levels which are in the middle of the energy gap between the valence and conduction bands. When energy is absorbed by an electron in the valence band it can cause it to jump to the recombination centre, that is provided the energy is large enough. Since the recombination centre is between the valence and conduction bands this is less energy than would be required for the jump from the valence band to the conduction band. The result of this jump to the recombination centre is that a hole is created in the valence band and an electron in the recombination centre. This electron may then receive additional thermal energy and end up in the conduction band. Thus an electron–hole pair is generated (Fig. 8.3(b)).

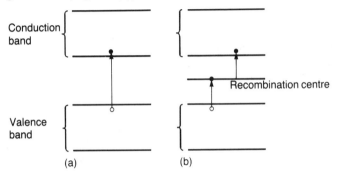

Fig. 8.3 Generation

(a)   (b)

Photoconductive cell

A *photoconductive cell* is a resistor whose resistance changes according to the intensity of light falling on it. Such resistors are semiconductors with the light providing the energy to increase the number of charge carriers and hence decrease the resistance.

The energy of a photon of light is given by $E = hf$, where h is Planck's constant and f the frequency of the light. Thus if the energy gap between the valence and conduction bands is E_g then the minimum frequency light can have in order to provide sufficient energy to move electrons across the gap and so generate electron–hole pairs is when

$$hf_{\text{min}} = E_g$$

This minimum frequency corresponds to a maximum wavelength λ_{max} where

$$c = f_{\text{min}}\lambda_{\text{max}}$$

and c is the speed of light. Thus the maximum wavelength that can be used is

$$\lambda_{\text{max}} = \frac{hc}{E_g} \qquad [2]$$

Cadmium sulphide is a commonly used material for such resistors when they are required to respond to visible light. It has an energy gap of 2.4 eV and thus the maximum wavelength possible with cadmium sulphide is 0.52 μm. The visible range of wavelengths of light is 0.4 to 0.7 μm, thus it would appear that cadmium sulphide will only respond to the blue end of the spectrum. However, cadmium sulphide is normally doped with atoms of other elements. Copper is used to give energy levels within the gap between the cadmium sulphide valence and conduction bands, i.e., recombination centres, such that only 1.7 eV is required for a photon to cause an electron to move from the valence band to this level and consequently produce an electron–hole pair. The maximum wavelength is thus increased to 0.7 μm, hence covering the entire visible spectrum. In addition to doping with copper the cadmium sulphide is doped with a hole-trap material such as chlorine. This produces hole traps just above the valence band. These can remove free holes and in doing so reduce the chance of the holes recombining with the free electrons. The result of this is that the free electrons have a longer lifetime.

For the detection of longer wavelengths either a semiconductor with a very small energy gap is required or one containing a dopant which gives energy levels not very far above the valence level. For detection of wavelengths in the near infra-red, i.e., wavelengths about 1 to 4 μm, compounds such as indium antimonide and lead sulphide can be used since they have energy gaps which are small enough. However the gaps are so small that thermal energy at 300 K results in large numbers of electrons moving from the valence band to the conduction band and can swamp the effect of illuminating the semiconductor. For this reason such photoconductive cells tend to be used at low temperatures. At 77 K indium antimonide has an energy gap of 0.23 eV and gives a maximum wavelength of 5.5 μm. Germanium doped with nickel or gold can be used for the near infra-red since the dopant introduces acceptor energy levels at about 0.2 eV above the valence band. Electrons can be excited from the valence band into these unoccupied levels and so leave holes in the valence band. Again, to avoid thermally excited electrons swamping the effect of the illumination, such photoconductive cells are used at low temperatures.

Example 1

Calculate the size of the energy gap between the valence and conduction bands in cadmium sulphide if the maximum wavelength which will result in a resistance change is 0.52 μm?
Planck's constant $h = 6.6 \times 10^{-34}$ J s
Speed of light $c = 3.0 \times 10^8$ m/s

Answer

Using equation [2]

$$\lambda_{max} = \frac{hc}{E_g}$$

hence

$$E_g = \frac{6.6 \times 10^{-34} \times 3.0 \times 10^8}{0.52 \times 10^{-6}}$$

$$= 3.8 \times 10^{-19} \text{ J} = 2.4 \text{ eV}$$

Example 2

Calculate the maximum wavelength which can be used to change the resistance of a photoconductive cell of germanium doped with gold if the gold introduces unoccupied energy levels 0.23 eV above the valence band?

Planck's constant $h = 6.6 \times 10^{-34}$ J s

Speed of light $c = 3.0 \times 10^8$ m/s

Answer

Using equation [2]

$$\lambda_{max} = \frac{hc}{E_g}$$

$$= \frac{6.6 \times 10^{-34} \times 3.0 \times 10^8}{0.23 \times 1.6 \times 10^{-19}} = 5.4 \ \mu\text{m}$$

Diffusion of charge carriers

The term *diffusion* is used to describe the process whereby particles spontaneously spread out with time, moving from areas where the particle concentration is high to regions where it is low. A simple illustration of this is when an open bottle of perfume or some other strongly smelling object is introduced into one corner of a room. After a while the perfume molecules will have spread throughout the room, i.e., diffusion will have occurred. In the context of this chapter we are concerned with an injection of charge carriers at some point in a material and the resulting diffusion of them into the material, i.e., the movement of the charge carriers from a position of high concentration to a region of lower concentration.

The mechanism responsible for diffusion, whether it be scent molecules or charge carriers, is the random motion of the particles. Such random motion inevitably leads to the particles spreading from a region of high concentration to one of low concentration. This can be illustrated by a simple model. Consider a box which is divided into two compartments (Fig. 8.4). In the left compartment there are six counters, numbered from 1 to 6. The right compartment is empty. When a six-sided die is thrown the number uppermost indicates that the 'random motion' of the indicated counter is such that it moves to the other compartment, the counters being able to move in either direction between the compartments. Thus if it is in the left it moves to the right and if in the right it moves to the left. The initial situation is of a high concentration of counters in the left compartment and zero in the right. After the 'game' has been played for a while an

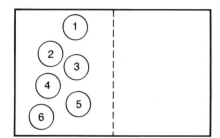

Fig. 8.4 A diffusion model

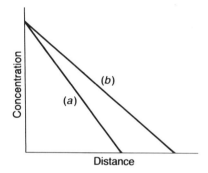

Fig. 8.5 Concentration gradients

equilibrium state is reached when, though particles are still moving from left to right and from right to left, there are roughly equal numbers of counters in each compartment. With larger numbers of counters the equality of numbers in the two compartments would be better. Thus, as a result of random motion the particles have spread from the high to low concentration region. Diffusion has occurred.

The rate at which particles diffuse across a plane is proportional to the concentration gradient (Fig. 8.5). This is known as *Fick's first law*. Thus the greater the difference in concentrations between two points the greater the rate of diffusion. Thus for Fig. 8.5, because the concentration gradient (*a*) is greater than that of (*b*) the rate of diffusion is greater for (*a*) than (*b*). The law can be expressed as

$$\text{rate of diffusion} = -D\frac{dC}{dx} \qquad [3]$$

where the rate of diffusion is the number of particles passing through unit area per second, dC/dx the concentration gradient and D a constant called the *diffusion coefficient*. The minus sign indicates that a positive flow of particles goes in the direction of falling concentration, i.e., a negative value of dC/dx.

Thus for the two-compartment model used above (Fig. 8.4) to represent diffusion, we can visualise the game involving in one case a start position with six counters in the left and none in the right compartments and in another a start with one hundred in the left and none in the right. The rate at which the counters move from left to right is greater initially with the hundred than with the six – the concentration gradient is greater.

Equation [2] represents the general form of the law that applies to all forms of diffusion. In the case of excess charge carriers being injected into a material, equation [2] can be written as

number of excess electrons diffusing through unit area per

$$\text{second} = -D_n\frac{dn}{dx}$$

for excess electrons, and for excess holes

number of excess holes diffusing through unit area per

$$\text{second} = -D_p\frac{dp}{dx}$$

with D_n being the *electron diffusion coefficient* and D_p the *hole diffusion coefficient*, dn/dx the excess electron concentration gradient and dp/dx the excess hole concentration gradient.

The motion of charge particles constitutes a current and thus the diffusion of excess electrons or excess holes constitutes a *diffusion current*. If the excess electrons each carry a charge $-q$ then the current passing through unit area, i.e., the current density J_n, is

$$J_n = -q \times \text{(number of excess electrons diffusing through unit area per second)}$$

Hence

$$J_n = qD_n\frac{dn}{dx} \qquad [4]$$

For excess hole movement the current density J_p is

$$J_p = +q \times \text{(number of excess holes diffusing through unit area per second)}$$

$$J_p = -qD_p\frac{dp}{dx} \qquad [5]$$

Example 3

What is the diffusion current density in a semiconductor if the hole density changes from 2.0×10^{10} holes/m^3 to 1.2×10^9 holes/m^3 in a distance of 12 μm? The hole diffusion coefficient for the material is 1.1×10^{-3} m^2/s and the charge carried by a hole is 1.6×10^{-19} C.

Answer

Using equation [5]

$$J_p = -qD_p\frac{dp}{dx}$$

$$= -1.6 \times 10^{-19} \times 1.1 \times 10^{-3} \times \left(\frac{2.0 \times 10^{10} - 1.2 \times 10^9}{12 \times 10^{-6}}\right)$$

$$= -2.8 \times 10^{-7} \text{ A/m}^2$$

Total current

A drift current occurs in a semiconductor if there is a potential difference, i.e., if there is an electric field. A diffusion current occurs if there is a carrier concentration gradient. The total current in a semiconductor when there is both an electric field and a carrier concentration gradient is the sum of these two terms. For electrons the drift current density J_n is (eqn [5] Ch. 7)

$$J_n = E\sigma$$

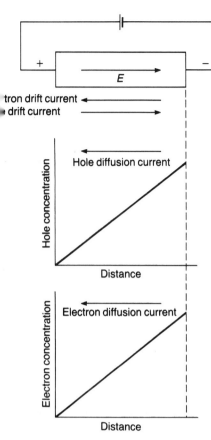

Fig. 8.6 Total current

where σ is the conductivity. This is (eqn [7] Ch. 7)

$$\sigma = nq\mu_n$$

where q is the charge carried by an electron, n the number of electrons per cubic metre and μ_n their mobility. Thus

$$J_n = q\mu_n En$$

The diffusion current density is given by equation [4] as

$$J_n = qD_n\frac{dn}{dx}$$

where D_n is the electron diffusion coefficient and dn/dx the electron concentration gradient. Thus for the electric field direction in the same direction as that of increasing carrier concentration, as in Fig. 8.6., the current density due to the electrons is

$$J_n = q\mu_n En + qD_n\frac{dn}{dx} \qquad [6]$$

The currents due to the drift and the diffusion are in the same direction in this case.

Similarly for holes, the total hole current density is

$$J_p = q\mu_p Ep - qD_p\frac{dp}{dx} \qquad [7]$$

where μ_p is the hole mobility, p the hole density, q the charge carried by a hole, D_p the hole diffusion coefficient and dp/dx the hole concentration gradient. The currents due to the hole drift current and the hole diffusion current are, in this case, in opposite directions.

Einstein's relationship

The free charge carriers in a material can be considered, in the absence of an electric field or concentration gradient, to be in random motion with a kinetic energy given by (see Ch. 7)

$$\tfrac{1}{2}mv^2 \approx kT$$

where v is the random velocity, k Boltzmann's constant and T the temperature on the kelvin scale. Thus the charge carriers can be thought of as bouncing about between the atoms in a completely chaotic manner with a random velocity which depends on the value of kT.

When an electric field is applied the charge carriers acquire, superimposed on their random velocities, a drift velocity. The relationship between the drift velocity and the mobility is given by (eqn [6] Ch. 7)

$$v = \mu E$$

with the mobility depending on the average time between collisions (see Ch. 7 and eqn [8]) and hence the random velocity.

When there is a concentration gradient the random motion of the charge carriers results in diffusion occurring and a motion, superimposed on their random velocities, of the charge carriers down the concentration gradient. The resulting current density, and hence superimposed diffusion velocity, is given for electrons by equation [4] as

$$J_n = qD_n \frac{dn}{dx}$$

where dn/dx is the concentration gradient and q the charge on each charge carrier. The diffusion coefficient D_n like the mobility depends on the collisions between the charge carriers and the atoms and hence the random velocity.

Einstein showed that the mobility and diffusion coefficient for a charge carrier are related, the relationship now being known as *Einstein's relationship*, by

$$\frac{\mu_n}{D_n} = \frac{\mu_p}{D_p} = \frac{q}{kT} \qquad [8]$$

Example 4

Calculate the electron and hole diffusion coefficients for silicon at 20 °C if the electron mobility is 0.15 m^2 V^{-1} s^{-1} and the hole mobility is 0.048 m^2 V^{-1} s^{-1}.
Boltzmann's constant = 1.38×10^{-23} J/K
Charge carried by holes and electrons = 1.6×10^{-19} C

Answer
Using equation [8]

$$\frac{\mu_n}{D_n} = \frac{\mu_p}{D_p} = \frac{q}{kT}$$

Hence

$$D_n = \frac{\mu_n}{(q/kT)} = \frac{0.15}{[1.6 \times 10^{-19}/(1.38 \times 10^{-23} \times 293)]}$$

$$= 3.8 \times 10^{-3} \text{ m}^2/\text{s}$$

$$D_p = \frac{\mu_p}{(q/kT)} = \frac{0.048}{[1.6 \times 10^{-19}/(1.38 \times 10^{-23} \times 293)]}$$

$$= 1.2 \times 10^{-4} \text{ m}^2/\text{s}$$

Excess carrier density profiles

So far in considering the diffusion of holes or electrons through a semiconductor the effect of any recombination of

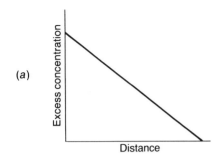

(a)

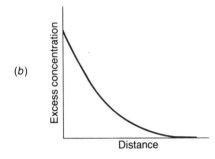

(b)

Fig. 8.7 Carrier density profiles, (a) with zero recombination, (b) with carrier recombination

these carriers has been ignored. This would be the situation if the diffusion takes place in much less than the average lifetime τ of the carriers. Then the excess carrier concentration profile, i.e., variation of carrier density with distance, looks like that shown in Fig. 8.7(a). If, however, there is recombination occurring then the excess carrier concentration profile looks more like that in Fig. 8.7(b).

Consider a thin layer of the semiconductor of thickness δx, as in Fig. 8.8. If the current density of excess electrons entering this layer is J_n then the recombination within this layer will reduce the current density to $(J_n - \delta J_n)$. If τ_n is the average lifetime of the excess electrons and there is initially an excess of n_e per cubic metre then the rate of recombination is given by (eqn [1])

$$\text{rate of recombination} = -\frac{n_e}{\tau_n}$$

Thus if the excess electrons take a time δt to traverse the distance δx then the number recombining in that time is

$$\text{number recombining} = -\frac{n_e \delta t}{\tau_n}$$

Current density is related to the number of charge carriers by $J = vnq$ (eqn [3] Ch. 7). Thus since the number of charge carriers on entering the layer is n_e then

$$J_n = v_e nq$$

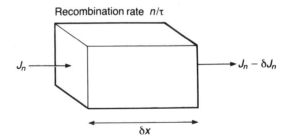

Recombination rate n/τ

J_n ⟶ ⟶ $J_n - \delta J_n$

δx

Fig. 8.8 Recombination

and since on leaving it the number of charge carriers has been reduced by those recombining

$$J_n - \delta J_n = vq\left(n_e - \frac{n_e \delta t}{\tau_n}\right)$$

Hence

$$\delta J_n = \frac{vqn_e \delta t}{\tau_n}$$

But $v = \delta x / \delta t$, hence

$$\delta J_n = \frac{q n_e \delta x}{\tau_n}$$

This can be expressed as the rate of change of current density with distance, dJ_n/dt, with

$$\frac{\mathrm{d}J_n}{\mathrm{d}x} = \frac{q n_e}{\tau_n} \qquad [9]$$

This equation is known as the *continuity equation*. A similar equation can be derived for the excess holes.

$$\frac{\mathrm{d}J_p}{\mathrm{d}x} = \frac{q p_e}{\tau_p} \qquad [10]$$

Electrons arrive at the layer (Fig. 8.8) as a result of a diffusion current given by equation [4] as

$$J_n = q D_n \frac{\mathrm{d}n}{\mathrm{d}x}$$

Thus using this equation to substitute for J_n on equation [9] gives

$$\frac{\mathrm{d}}{\mathrm{d}x} \left(q D_n \frac{\mathrm{d}n}{\mathrm{d}x} \right) = \frac{q n_e}{\tau_n}$$

$$\frac{\mathrm{d}^2 n}{\mathrm{d}x^2} = \frac{n_e}{D_n \tau_n}$$

If $n_e = n_{e0}$ at $x = 0$ and n_e dies down to zero when x reaches infinity then the solution of this differential equation is

$$n_e = n_{e0} \exp\left(- \frac{x}{\sqrt{(D_n \tau_n)}} \right)$$

The quantity $\sqrt{(D_n \tau_n)}$ is called the *electron diffusion length* L_n. Thus

$$L_n = \sqrt{(D_n \tau_n)} \qquad [11]$$

Hence

$$n_e = n_{e0} \exp\left(-x/L_n\right) \qquad [12]$$

Figure 8.9 shows how the number of excess charge carriers varies with distance for this exponential relationship.

When $x = L_n$ then equation [12] gives

$$n_e = n_{e0} \exp(-1) = 0.37\, n_{e0}$$

The diffusion length is the distance from $x = 0$ at which the number of excess charge carriers has dropped to 0.37 of its initial value.

Similar relationships can be derived for excess holes, i.e.,

$$L_p = \surd(D_p\tau_p) \tag{13}$$

$$p_e = p_{e0} \exp(-x/L_p) \tag{14}$$

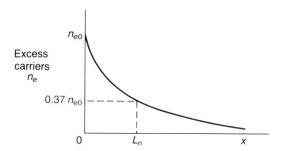

Fig. 8.9 Excess carrier density profile

Example 5

What are the diffusion lengths for excess holes and electrons if both have mean lifetimes of 1.0×10^{-5} s? $D_n = 3.8 \times 10^{-3}$ m^2/s and $D_p = 1.3 \times 10^{-3}$ m^2/s.

Answer

Using equation [11]

$$L_n = \surd(D_n\tau_n) = \surd(3.8 \times 10^{-3} \times 1.0 \times 10^{-5})$$
$$= 1.9 \times 10^{-4} \text{ m} = 0.19 \text{ mm}$$

$$L_p = \surd(1.3 \times 10^{-3} \times 1.0 \times 10^{-5})$$
$$= 1.1 \times 10^{-4} \text{ m} = 0.11 \text{ mm}$$

Contacts between metals

Consider two metals with different Fermi levels put in contact, as in Fig. 8.10. The situation is rather like putting in contact two cans of water in which the water level in one is higher than the other. Water will flow between the cans until the levels are the same in each can. The same happens with the electrons in the two metals. Electrons diffuse across the boundary until the Fermi level is the same on both sides of the junction. This happens because the probability of finding an electron must become the same on both sides since the electrons are able to move freely across the boundary.

An alternative way of looking at the situation is that metal A has more occupied energy levels and so more electrons per cubic metre than metal B. There is thus a concentration gradient across the boundary and hence diffusion will occur.

Initially both metals had no net charge, but the metal from which there has been a net flow of electrons in this diffusion process must end up with a positive charge and the metal to which there is a net input of electrons a negative charge. The electrons thus flow into a metal which is becoming more and

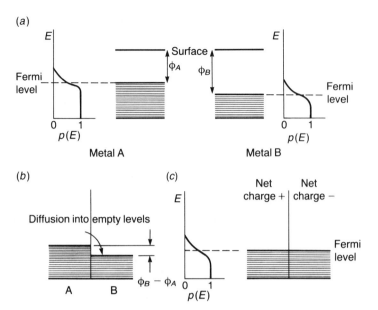

Fig. 8.10 Contact between two metals, (a) before contact, (b) on contact, (c) after diffusion

more negatively charged as the flow proceeds and so repulsive forces are experienced by the electrons which increase until a point is reached when they are large enough to stop further electrons crossing. An electric field opposing the movement of the electrons builds up. Since an electric field means a potential gradient, a potential difference is produced between the two sides of the junction. This potential difference is called a *contact potential difference*.

If the initial energy of the Fermi level in metal A was E_{FA} and that in metal B E_{FB}, then to move an electron of charge q between these levels requires an energy of $(E_{FA} - E_{FB})$ and so the potential difference V_{AB} is given by equation [3], Chapter 2.

$$E_{FA} - E_{FB} = qV_{AB} \qquad [15]$$

Since charge is moved, as a result of diffusion, between the two metals until the Fermi level is the same on both sides the energy that must be expended is qV_{AB}. Thus V_{AB} is the contact potential.

The position of the Fermi level for a material can be specified relative to the energy required to remove an electron from the material (see Fig. 8.10). This is the work function ϕ (see Chs 6 and 7). Thus the initial difference in the energies of the Fermi levels for the two metals is $(\phi_B - \phi_A)$, where ϕ_A and ϕ_B are the work functions of the two metals. Hence

$$\phi_B - \phi_A = qV_{AB} \qquad [16]$$

It is customary to state work function values in electron volts (eV). To convert the work function values in electron volts to joules we need to multiply them by the charge on the electron q. Thus with the work functions in electron volts

$$qV_{AB} = q\phi_B - q\phi_A$$

$$V_{AB} = \phi_B - \phi_A \tag{17}$$

Example 6

What is the contact potential difference at a junction between aluminium and copper if aluminium has a work function of 4.2 eV and copper 4.5 eV?

Answer

Using equation [17]

$$V_{AB} = \phi_B - \phi_A = 4.5 - 4.2 = 0.3 \text{ V}$$

The p–n junction

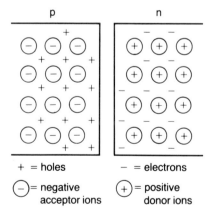

p   n

+ = holes   − = electrons

$\ominus$ = negative acceptor ions $\oplus$ = positive donor ions

Fig. 8.11 Charges in p and n materials

In p-type semiconductor material the dopant atoms are acceptors. At the location of a dopant atom a bonding electron is missing. This location is electrically neutral. However, a valence electron from one of the semiconductor atoms can move into the missing bond position. The result of this is that the dopant atom has now a net negative charge and is said to be ionised. A consequence of this movement of the valence electron is that a hole has been left in the bond of the semiconductor atoms and that atom in being one electron short has a net positive charge. We can consider the hole effectively to have a positive charge. The number of ionised dopant atoms N_a equals the number of holes p. The charges in such a material can be pictorially represented by the left-hand picture in Fig. 8.11.

In n-type material the dopant atoms are donors. At the location of a dopant atom there is an extra electron over and above those required for bonding. This location is electrically neutral. However, this electron may become detached, with the result that the donor atom now has a net positive charge, i.e., is a positive ion, and there is a free electron. The number of ionised donor atoms N_d equals the number of free electrons n. The charges in such a material can be pictorially represented by the right-hand picture in Fig. 8.11.

When the p and n versions of a semiconductor are put in contact to form a p–n junction, there is a higher concentration of holes in the p material than in the n material and a higher concentration of free electrons in the n material than the p material. We thus have a concentration gradient for holes to diffuse from the p to the n materials and another concentration gradient for electrons to diffuse from the n to p materials.

Before any diffusion takes place the number of free charge carriers in each of the materials is represented by the graph in Fig. 8.12(a). For the p material the number of holes per cubic metre p equals the number of ionised acceptor atoms N_A. The number of free electrons in the p material is given by equation [1], Chapter 7,

$$np = n_i^2$$

where n_i is the intrinsic carrier concentration. Thus n is n_i^2/p and so n_i^2/N_a. For the n material the number of free electrons per cubic metre n equals the number of ionised donor atoms N_d. Using the above equation, the number of holes in the n material is thus n_i^2/N_d.

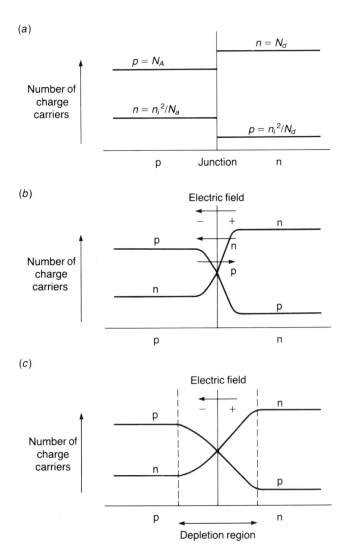

Fig. 8.12 Charge carriers, (a) before diffusion, (b) during diffusion, (c) at equilibrium

When diffusion occurs there is a net flow of holes from the p side of the junction to the n side and a net flow of electrons from the n side to the p side, as illustrated in Fig. 8.12(*b*). The electrons that have diffused across the junction will eventually meet up with holes and recombine, likewise the holes that have diffused across the junction will recombine with electrons. But before this diffusion occurred, each side of the junction was electrically neutral. The diffusion thus leads to the p side becoming negatively charged, because it is left with 'unbalanced' negative ions, and the n side positively charged, because it is left with 'unbalanced' negative ions. The result is an electric field across the junction. As the diffusion continues more and more of the dopant ions on either side of the junction become depleted of their 'balancing' charges and so the charges build up on each side and the electric field increases. The electric field, however, is in such a direction as to oppose the build up of charge and so eventually a situation is reached when the electric field is large enough to bring the net movement of charges across the junction to an end. We can think of the hole diffusion current across the junction being balanced by a hole drift current which has resulted from there being an electric field. Similarly the electron diffusion current is balanced by the electron drift current which has resulted from there being an electric field. The equilibrium situation thus occurs when the tendency of the carriers to diffuse is balanced by their tendency to drift in the electric field. Thus at equilibrium there is an electric field across the junction and hence a potential difference. The term *depletion region* is used to describe the region on both sides of the junction in which the dopant atoms have been depleted of their 'balancing' charges.

The p–n junction at equilibrium

The current density across the junction resulting from electron movement is the sum of that due to diffusion and drift, hence according to equation [6]

$$J_n = q\mu_n En + qD_n\frac{dn}{dx}$$

where q is the charge carried by an electron, μ_n its mobility, E the electric field, n the number of electrons per cubic metre, D_n the electron diffusion coefficient and dn/dx the concentration gradient. At equilibrium there is no current and so

$$0 = q\mu_n En + qD_n\frac{dn}{dx}$$

Hence

$$E \, dx = - \frac{D_n}{\mu_n} \frac{1}{n} dn$$

Since the electric field is the potential gradient, i.e. $E = -dV/dx$ (eqn [2], Ch. 2), then the potential difference V_{pn} across the depletion layer is

$$V_{pn} = - \int_n^p E \, dx$$

with the integration being taken from a point far enough on one side of the junction in the p material for the effects of the junction to be negligible to a corresponding point on the other side in the n material. Thus, combining the two equations,

$$V_{pn} = \int_n^p \frac{D_n}{\mu_n} \frac{1}{n} dn$$

On the p side the number of electrons is n_i^2/N_a and this changes to N_d on the n side. Thus integrating between these limits gives

$$V_{pn} = \frac{D_n}{\mu_n} \ln \left(\frac{n_i^2}{N_a N_d} \right) \qquad [18]$$

Similarly for holes, the total hole current density is given by equation [7] as

$$J_p = q\mu_p E p - q D_p \frac{dp}{dx}$$

where μ_p is the hole mobility, p the hole density, q the charge carried by a hole, D_p the hole diffusion coefficient and dp/dx the hole concentration gradient. At equilibrium there is no current, hence an equation similar to that derived above for electrons can be obtained.

$$V_{pn} = \frac{D_p}{\mu_p} \ln \left(\frac{n_i^2}{N_a N_d} \right) \qquad [19]$$

Since both these equations, [18] and [19], refer to the same potential difference then D_n/μ_n must equal D_p/μ_p. This is in fact part of Einstein's relation (eqn [8]) with both these quantities equalling kT/q. Thus the equations can be written as

$$V_{pn} = \frac{kT}{q} \ln \left(\frac{n_i^2}{N_a N_d} \right) \qquad [20]$$

The potential difference across the junction thus depends on the temperature.

Example 7

A silicon p–n junction has a density of 1.0×10^{22} acceptor atoms per cubic metre on the p side and 4.0×10^{21} donor atoms per cubic metre on the n side. If the intrinsic concentration n_i is 1.6×10^{16} /m³ at 300 K, what is the potential difference across the junction at this temperature?

Boltzmann's constant = 1.38×10^{-23} J/K

Charge carried by a carrier = 1.6×10^{-19} C

Answer

Using equation [20] and assuming that all the acceptor and donor atoms are ionised,

$$V_{pn} = \frac{kT}{q} \ln \left(\frac{n_i^2}{N_a N_d} \right)$$

$$= \frac{1.38 \times 10^{-23} \times 300}{1.6 \times 10^{-19}} \ln \left(\frac{(1.6 \times 10^{16})^2}{1.0 \times 10^{22} \times 4.0 \times 10^{21}} \right)$$

$$= -0.67 \text{ V}$$

The minus sign indicates that the potential difference is in the opposite direction, i.e., from n to p.

Carrier densities for the p–n junction

At equilibrium there is an internal, or contact, potential difference V_{pn} at a p–n junction. This is given by equation [20] as

$$V_{pn} = \frac{kT}{q} \ln \left(\frac{n_i^2}{N_a N_d} \right)$$

This equation can be rearranged to give

$$\frac{n_i^2}{N_d} = N_a \exp (qV_{pn}/kT)$$

Thus if we consider the electron and hole carrier densities n_{n0} and p_{n0} at equilibrium in the n material, outside the junction depletion layer, then since the electron carrier density n_{n0} is generally about the same as the density of donor atoms N_d and the hole carrier density p_{n0} is given by n_i^2/n_{n0}

$$\frac{n_i^2}{n_{n0}} = p_{n0} = N_a \exp (qV_{pn}/kT)$$

The subscript 0 is used to indicate that the carrier densities are equilibrium ones. But N_a is the hole carrier density p_{p0} in the p material, outside the depletion layer. Thus

$$p_{n0} = p_{p0} \exp (qV_{pn}/kT)$$

Since the potential difference in going from p to n is minus that in going from n to p, i.e., $V_{pn} = - V_{np}$, the equation is more usually written as

$$p_{n0} = p_{p0} \exp\left(- qV_{np}/kT\right) \qquad [21]$$

Similarly an equation can be developed for the electrons

$$n_{p0} = n_{n0} \exp\left(- qV_{np}/kT\right) \qquad [22]$$

The effect of bias for the p–n junction

A p–n junction at equilibrium has a built-in, or contact, potential difference with the n side at a positive potential with respect to the n side of the junction. When an external potential difference is applied to the junction then *bias* is said to have been applied. The bias is said to be *forward bias* when it is in the opposite direction to the built-in potential difference and so decreases the potential difference across the junction, i.e., when the p side of the junction is connected to the positive terminal of the voltage supply, and *reverse bias* when it is in the same direction and increases the potential difference across the junction, i.e., when the n side of the junction is connected to the positive terminal of the voltage supply. Thus if the applied potential difference is V, then with forward bias the total potential difference across the junction becomes $(V_{np} - V)$ and with reverse bias $(V_{np} + V)$.

With forward bias equation [21] can be written as

$$p_{n1} = p_{p1} \exp\left[- q(V_{np} - V)/kT\right]$$

and equation [22] as

$$n_{p1} = n_{n1} \exp\left[- q(V_{np} - V)/kT\right]$$

with the subscript 1 indicating that the carrier densities are not the equilibrium values, subscript 0, but the values that occur with this bias. The hole density at equilibrium in the p material, away from the depletion layer, is very large and the bias makes little difference in its value. Thus p_{p1} is effectively p_{p0}. Similarly the electron density at equilibrium in the n material is very large and the bias makes little difference in its value. Thus n_{n1} is effectively n_{n0}. Thus the two equations can be written as

$$p_{n1} = p_{p0} \exp\left[- q(V_{np} - V)/kT\right]$$
$$= p_{p0} \exp\left[-qV_{np}/kT\right] \times \exp\left[qV/kT\right]$$

and substituting using equation [21]

$$p_{n1} = p_{n0} \exp\left[qV/kT\right] \qquad [23]$$

Likewise

$$n_{p1} = n_{n0} \exp\left[- q(V_{np} - V)/kT\right]$$

$$= n_{n0} \exp\left[-qV_{np}/kT\right] \times \exp\left[qV/kT\right]$$

and substituting using equation [22]

$$n_{p1} = n_{p0} \exp\left[qV/kT\right] \qquad [24]$$

Equations [23] and [24] state how the number of minority charge carriers in the n and in the p materials change as a result of the applied bias.

With forward bias V is a positive quantity and so the exponential term is greater than 1. Hence n_{p1} is greater than n_{p0} and p_{n1} is greater than p_{n0}. Thus a forward bias has increased the number of minority charge carriers in the p and n material on either side of the depletion layer.

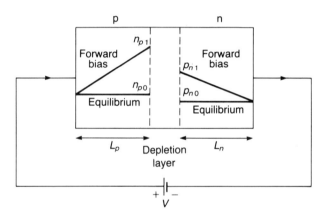

Fig. 8.13 Minority carrier densities in a forward-biased p–n junction

The effect of such changes is to produce concentration gradients for the minority charge carriers in the material on either side of the depletion layer (Fig. 8.13). At each end of the piece of material it is assumed there is an ohmic contact (see later in this chapter) with metal wires connecting it to the source of bias voltage. Each such ohmic contact keeps the carrier densities down at their equilibrium values, as a result of recombinations with charge carriers in excess of those values. The hole concentration gradient in the n material is thus $(p_{n1} - p_{n0})/L_n$, where L_n is the length of the n material. Hence the current density due to hole movement in the n material, J_p, is given by equation [5] as

$$J_p = -qD_p\frac{dp}{dx} = -qD_p\frac{(p_{n1} - p_{n0})}{L_n}$$

Substituting from p_{n1} using equation [23] gives

$$J_p = -\frac{qD_p p_{n0}}{L_n}\left[\exp\left(qV/kT\right) - 1\right]$$

Similarly the current density due to the electron movement in the p material, J_n, is given by equation [4] as

$$J_n = qD_n\frac{dn}{dx} = qD_n\frac{(n_{p1} - n_{p0})}{L_p}$$

where L_p is the length of the p material and using equation [24]

$$J_n = \frac{qD_n n_{p0}}{L_p}[\exp{(qV/kT)} - 1]$$

The current due to the minority electrons in the p material is just the same as the current due to the majority electrons in the n material. Thus the total current in the n material is the sum of that due to the minority holes and the majority electrons, the current densities being added since the two are of opposite charges flowing in opposite directions. Thus

$$\text{total } J = \frac{qD_p p_{n0}}{L_p}[\exp(qV/kT)] + \frac{qD_n n_{p0}}{L_n}[\exp(qV/kT)]$$

Since the current I is the current density multiplied by the cross-sectional area, then

$$I = Aq\left(\frac{D_p p_{n0}}{L_p} + \frac{D_n n_{p0}}{L_n}\right)[\exp{(qV/kT)} - 1]$$

All the constants for the junction can be combined as I_0 and so

$$I = I_0[\exp{(qV/kT)} - 1] \qquad [25]$$

This is the *characteristic diode equation*.

Figure 8.14 shows a graph of it and so how the current I varies the applied potential difference V. With forward bias V is positive and so as the exponential term increases the current I increases as V increases. With reverse bias V is negative and so as V becomes more negative the exponential term becomes smaller and smaller. Thus when the exponential term has become insignificant the current becomes equal to $-I_0$. The current I_0 is often called the *saturation current*.

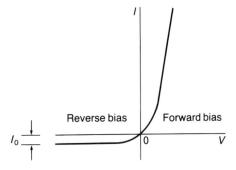

Reverse bias   Forward bias

Fig. 8.14 The p–n junction current–voltage characteristic

Example 8

A germanium p–n junction diode at 300 K gives a current of 1.0 mA when forward biased by 0.20 V. What is the saturation current for the diode?
Boltzmann's constant $k = 1.38 \times 10^{-23}$ J/K
Charge carried by a charge carrier $= 1.6 \times 10^{-19}$ C

Answer

Using equation [25]

$$I = I_0[\exp{(qV/kT)} - 1]$$

$$\left(\frac{I}{I_0}\right) = \exp{\left(\frac{qV}{kT}\right)} - 1 = \exp{\left(\frac{1.6 \times 10^{-19} \times 0.20}{1.38 \times 10^{-23} \times 300}\right)} - 1$$

Hence with $I = 1.0$ mA

$$I_0 = 4.4 \times 10^{-7} \text{ A}$$

Width of depletion region

In the depletion region of a p–n junction we have essentially just the ionised dopant atoms, with ionised acceptor atoms on the p side and ionised donor atoms on the n side (Fig. 8.15). The type of junction shown in the figure is called an *abrupt junction* since there is an abrupt change from p material to n material (another form of junction is a graded one with there being a gradual transition from one type of impurity to the other). If there are N_a ionised acceptor atoms per cubic metre, each with a charge q, then in a layer of thickness δx and area A there will be a charge of $N_a q A \delta x$ (Fig. 8.16). But the electric field E due to a charge density σ is given by equation [8] Chapter 2,

$$\sigma = \varepsilon E$$

where the charge density σ is charge per unit area (eqn [7] Ch. 2) and ε the permittivity (see eqn [6] Ch. 2). Thus the electric field δE produced by the thin layer δx is

$$\sigma = \frac{N_a q \delta x}{\varepsilon}$$

Thus to find the electric field at any distance x from the edge of the p depletion region what is required is the summation of all the effects of the δx layers across the width of the p depletion region up to distance x, i.e., an integration between $-x_p$ and $-x$ (the minus signs are because the zero position of

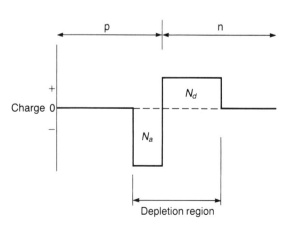

Fig. 8.15 Charges in the depletion region

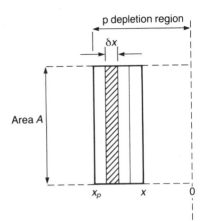

p depletion region

δx

Area A

x_p x 0

Fig. 8.16 Charges on layers in depletion region

x has been taken as in the depletion region at the junction between the p depleted region and the n depleted region (see Fig. 8.15)). Thus

$$E = \int_{-x_p}^{-x} \frac{N_a q}{\varepsilon} \, dx$$

$$E = \frac{N_a q}{\varepsilon} (x_p - x)$$

The electric field is the potential gradient, $E = -\,dV/dx$ (eqn [2] Ch. 2). Thus the potential difference across the p depletion region is the integral of $-E dx$ between the limits of $x = 0$ and $x = -x_p$.

$$V_p = -\int_0^{-x^p} \frac{N_a q}{\varepsilon} (x_p - x) \, dx$$

$$= -\frac{N_a q x_p^2}{2\varepsilon}$$

A similar equation can be developed for the potential difference across the n depletion region.

$$V_n = -\frac{N_d q x_n^2}{2\varepsilon}$$

where N_d is the number of ionised donor atoms per cubic metre and x_n the width of the n depletion region.

If V is the total potential difference across the depletion region then

$$V = V_p + V_n = -\left(\frac{N_a q x_p^2}{2\varepsilon} + \frac{N_d q x_n^2}{2\varepsilon} \right)$$

$$= -\frac{q}{2\varepsilon}(N_a x_p^2 + N_d x_n^2) \qquad [26]$$

The total charge in the p depleted region is the same as the total charge in the n depleted region. Thus since there are N_a ionised acceptors per cubic metre, each with charge q, in the p depletion region and the region has a thickness of x_p and area A

charge in p depletion region $= N_a q A x_p$

Similarly

charge in n depletion region $= N_d q A x_n$

Hence

$$N_a q A x_p = N d_q A x_n$$

and so

$$N_a x_p = N_d x_n$$

Thus $x_p = N_d x_n / N_a$, and substituting this for x_p in equation [26] gives

$$V = -\frac{q}{2\varepsilon}\left(\frac{N_d^2 x_n^2}{N_a} + N_d x_n^2\right)$$

and so

$$x_n = \frac{1}{N_d}\sqrt{\left[\frac{2\varepsilon(-V)}{q(1/N_a + 1/N_n)}\right]} \tag{27}$$

Similarly, substituting for x_n in equation [26] gives

$$V = -\frac{q}{2\varepsilon}\left(N_a x_p^2 + \frac{N_p^2 x_p^2}{N_n}\right)$$

and

$$x_p = \frac{1}{N_a}\sqrt{\left[\frac{2\varepsilon(-V)}{q(1/N_a + 1/N_d)}\right]} \tag{28}$$

The total width w of the depletion region is $(x_n + x_p)$, hence

$$w = \left(\frac{1}{N_d} + \frac{1}{N_a}\right)\sqrt{\left[\frac{2\varepsilon(-V)}{q(1/N_a + 1/N_d)}\right]}$$

$$w = \sqrt{\left[\frac{2\varepsilon}{q}(-V)(1/N_a + 1/N_d)\right]} \tag{29}$$

The width of the depletion region thus depends on the amount of doping on each side of the junction and the potential difference across the junction. The higher the doping the thinner is the depletion region. For a junction between a highly doped material and low-doped material, the depletion region will be mainly on the low-doped side of the junction. If there is forward bias and the applied potential difference is in the opposite direction to the built-in potential difference, and so V is made smaller, then the width is reduced. If there is reverse bias then V is made larger and so the width increases.

Example 9

An abrupt silicon p–n junction has an acceptor density of 3.0×10^{23} /m^3 and a donor density of 3.0×10^{22} /m^3. What is the width of the depletion region at 300 K when the junction has (*a*) a reverse bias of 10 V, (*b*) a forward bias of 0.50 V?

The relative permittivity of silicon is 12.
Permittivity of free space $= 8.9 \times 10^{-12}$ F/m.
Boltzmann's constant $= 1.38 \times 10^{-23}$ J/K.
Intrinsic concentration for silicon at 300 K $= 1.4 \times 10^{16}$ /m^3.
Charge carried by an electron or hole $= 1.6 \times 10^{-19}$ C.

Answer

The built-in potential difference across the junction can be obtained by the use of equation [20].

$$V_{pn} = \frac{kT}{q} \ln\left(\frac{n_i^2}{N_a N_d}\right)$$

$$= \frac{1.38 \times 10^{-23} \times 300}{1.6 \times 10^{-19}} \ln\left[\frac{(1.4 \times 10^{16})^2}{3.0 \times 10^{22} \times 3.0 \times 10^{23}}\right]$$

$$= -0.81 \text{ V}$$

The width of the depletion layer is found using equation [29]

$$w = \sqrt{\left[\frac{2\varepsilon}{q}(-V)(1/N_a + 1/N_d)\right]}$$

(a) With the reverse bias of 10 V then $V = -10 - 0.81 = -10.81$ V. Thus

$$w = \sqrt{\left[\frac{2 \times 12 \times 8.9 \times 10^{-12}}{1.6 \times 10^{-19}} \times 10.81\left(\frac{1}{3.0 \times 10^{22}} + \frac{1}{3.0 \times 10^{23}}\right)\right]}$$

$$= 7.3 \times 10^{-7} \text{ m}$$

(b) With the forward bias of 0.5 V then $V = 0.50 - 0.81 = -0.31$ V. Thus

$$w = \sqrt{\left[\frac{2 \times 12 \times 8.9 \times 10^{-12}}{1.6 \times 10^{-19}} \times 0.31\left(\frac{1}{3.0 \times 10^{22}} + \frac{1}{3.0 \times 10^{23}}\right)\right]}$$

$$= 1.2 \times 10^{-7} \text{ m}$$

Example 10

An abrupt silicon p–n junction has an acceptor density of 1.0×10^{23} /m^3 and a donor density of 1.0×10^{20} /m^3. What is the width of the depletion region at 300 K when the junction has a reverse bias of 10 V? What is depletion region thickness if the donor density is increased to 1.0×10^{21} m^3?
The relative permittivity of silicon is 12.
Permittivity of free space = 8.9×10^{-12} F/m.
Boltzmann's constant = 1.38×10^{-23} J/K.
Intrinsic concentration for silicon at 300 K = 1.4×10^{16} /m^3.
Charge carried by an electron or hole = 1.6×10^{-19} C.

Answer

The built-in potential difference across the junction can be obtained by the use of equation [20].

$$V_{pn} = \frac{kT}{q} \ln\left(\frac{n_i^2}{N_a N_d}\right)$$

$$= \frac{1.38 \times 10^{-23} \times 300}{1.6 \times 10^{-19}} \ln\left[\frac{(1.4 \times 10^{16})^2}{1.0 \times 10^{20} \times 1.0 \times 10^{23}}\right]$$

$$= -0.64 \text{ V}$$

The width of the depletion layer is found using equation [29]

$$w = \sqrt{\left[\frac{2\varepsilon}{q}(-V)(1/N_a + 1/N_d) \right]}$$

With the reverse bias of 10 V then $V = -10 - 0.64 = -10.64$ V. Thus

$$w = \sqrt{\left[\frac{2 \times 12 \times 8.9 \times 10^{-12}}{1.6 \times 10^{-19}} \times 10.64 \left(\frac{1}{1.0 \times 10^{20}} + \frac{1}{1.0 \times 10^{23}} \right) \right]}$$

$$= 1.2 \times 10^{-5} \text{ m}$$

With the changed doping

$$V_{pn} = \frac{1.38 \times 10^{-23} \times 300}{1.6 \times 10^{-19}} \ln \left[\frac{(1.4 \times 10^{16})^2}{1.0 \times 10^{21} \times 1.0 \times 10^{23}} \right]$$

$$= -1.1 \text{ V}$$

With the reverse bias of 10 V then $V = -10 - 1.1 = -11.1$ V. Thus

$$w = \sqrt{\left[\frac{2 \times 12 \times 8.9 \times 10^{-12}}{1.6 \times 10^{-19}} \times 11.1 \left(\frac{1}{1.0 \times 10^{21}} + \frac{1}{1.0 \times 10^{23}} \right) \right]}$$

$$= 3.9 \times 10^{-6} \text{ m}$$

Increasing the doping has reduced the depletion region width.

The p–n junction and energy bands

Figure 8.17(a) shows the energy bands of separate p and n forms of a semiconductor. The n form has many conduction electrons, the p form many holes. With the p form the Fermi level is in the lower half of the energy gap between the conduction and valence bands, with the n form it is in the upper half. Because the p and n materials are both doped versions of the same semiconducting material the positions of the conduction and valence bands are the same for both. When there is a junction between the p and n forms of a semiconductor electrons spill across (diffuse) from the n side to the p side and holes from the p side to the n side (Fig. 8.17(b)). Think of electrons as being like ball bearings which will move downwards given the chance, and holes as bubbles which will rise if given the chance. When there is equilibrium the Fermi level must be the same throughout. For this to happen the energy bands of the n material must be moved downwards relative to the n material to give the picture shown in Fig. 8.17(c).

Between the n and p materials, i.e., in the depletion region, the energy levels slope. A horizontal energy level indicates that the energy of a charge in that level does not vary with distance. A sloping energy level indicates that the energy of a charge in that level varies with distance. Since the energy needed to move a unit charge between two points is the

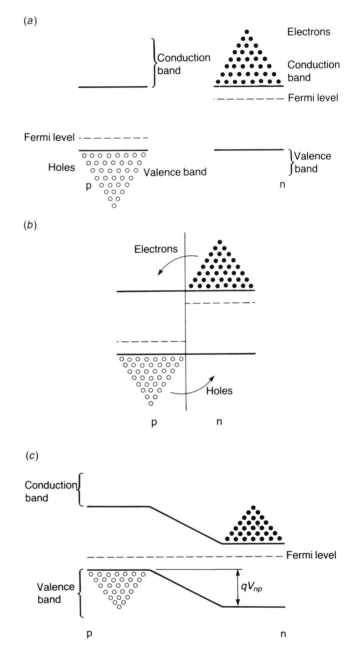

Fig. 8.17 Energy bands, (a) p and n separate, (b) p and n in contact, (c) p–n junction in equilibrium

potential difference (potential difference = energy/charge, eqn [3] Ch. 2), then a sloping energy level between two points indicates that there is a potential difference between them, i.e., there is a potential gradient. But the presence of a potential gradient indicates that there is an electric field

($E = - \mathrm{d}V/\mathrm{d}x$, eqn [2] Ch. 2), the direction of the electric field being in the opposite direction to that of increasing potential and so increasing energy. Thus in the absence of an electric field the energy bands are horizontal but when there is an electric field the energy bands slope, the direction of the electric field being up the slope. Thus in Fig. 8.17 the energy bands slope across the depletion region because there is an electric field, the direction of the electric field being from n to p. Since the direction of an electric field is the direction of the force on a positive charge, with electrons having a negative charge they will move in the opposite direction to the field and so drift down a sloping energy level. Holes, having a positive charge, move in the direction of the field and so move up a sloping energy level. It is for this reason that we can think of electrons as being like ball bearings which will move downwards given the chance, i.e., given an energy level sloping downwards, and holes as bubbles that will rise if given the chance, i.e., an energy level sloping upwards. The resulting energy difference between the energy bands at equilibrium is qV_{np}, where V_{np} is the potential difference produced across the junction.

When the p–n junction is reverse biased by a potential difference V, i.e., the p side is made more positive than the n side by a potential difference V, the total potential difference across the junction becomes $(V + V_{np})$ and the energy band picture becomes as shown in Fig. 8.18(a). When the junction is forward biased, i.e., the p side is made more negative by a potential difference V, the total potential difference across the junction becomes $(V_{np} - V)$ and the picture becomes as shown in Fig. 8.18(c). Biasing produces a difference in the Fermi levels on each side of the junction.

With the change from equilibrium to forward bias the number of electrons that can make it from the n side, where there are many conduction electrons, to the p side of the junction, where there are few, is increased (Fig. 8.18(c)). This is because the energy which electrons in the conduction band on the n side have to have to get into the p side conduction band is reduced, the energy hill to be surmounted is smaller. The change also increases the number of holes that can move from the p side to the n side. The net result is a current.

With the change from equilibrium to reverse bias the energy needed by the conduction band electrons on the n side to get into the conduction band on the p side has been increased, the energy hill to be surmounted is bigger (Fig. 8.18(a)). The change also does not increase the number of holes that can move from the p side to the n side. Hence there is no current produced by reverse biasing.

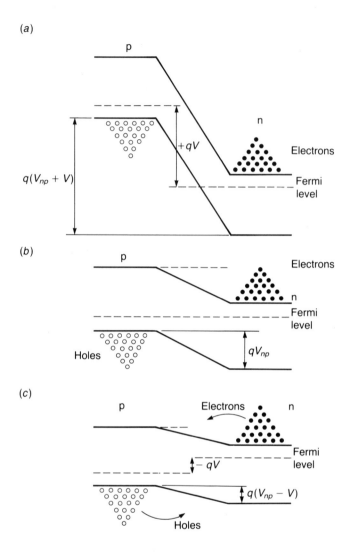

Fig. 8.18 Energy bands, (a) reverse bias, (b) equilibrium, (c) forward bias

Example 11

A germanium p–n diode has, at equilibrium, a contact potential of 0.40 V. Then a potential difference of 2.0 V is applied across the junction to reverse bias it. What is the energy difference between the bottom levels of the conduction bands in the p- and n-type materials at (a) equilibrium, (b) when reverse biased?

Charge carried by an electron = 1.6×10^{-19} C

Answer

(a) At equilibrium the energy difference between the bottom levels of the conduction bands is qV_{np}, where V_{np} is the contact potential. Thus the energy difference is

$$\text{energy difference} = 1.6 \times 10^{-19} \times 0.40 = 6.4 \times 10^{-20} \text{ J}$$

If the energy is expressed in units of electron volts then the energy difference is 0.40 eV.

(b) With reverse bias the energy difference between the bottom levels of the conduction bands is $q(V_{np} + V)$, i.e.

energy difference $= 1.6 \times 10^{-19} (0.40 + 2.0) = 3.84 \times 10^{-19}$ J

If the energy is expressed in electron volts it is 2.40 eV.

Metal–semiconductor contacts

When a junction is formed between a metal and a semiconductor, at equilibrium the Fermi levels in the two materials will become aligned. The consequences of this alignment depend on the relative values of the metal work function and the semiconductor work function, the work function being the energy needed for an electron at the Fermi level to escape from the material.

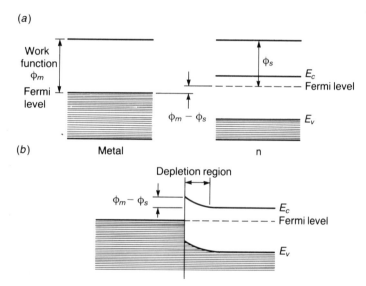

Fig. 8.19 Metal and n-type semiconductor junction when $\phi_m > \phi_s$

For a metal to n-type semiconductor junction where the metal work function ϕ_m is greater than the semiconductor work function ϕ_s the resulting energy band picture at equilibrium is as shown in Fig. 8.19(b). To align the Fermi levels the energy bands of the semiconductor have had to be shifted downwards relative to those of the metal. Electrons diffuse from the semiconductor to the metal with the result that the metal becomes negatively charged. A depletion region is produced in the semiconductor where electrons leave positive ions behind and so that side of the junction is positively charged. This leads to an electric field developing and, as earlier discussed for the p–n junction, equilibrium occurs when the diffusion current is balanced by the drift current resulting from the electric field. Because there is an electric field across the

(a)

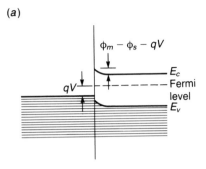

(b)

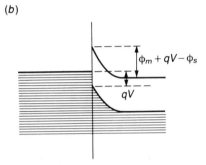

Fig. 8.20 Rectifying metal and n-type semiconductor junction with (a) forward bias, (b) reverse bias

junction there is a potential difference between the two sides of the junction. At equilibrium, when there is no net movement of charge across the junction, the energy barrier over which electrons from the semiconductor band have to climb to fall into the metal conduction levels is $(\phi_m - \phi_s)$ and so the contact potential difference is then $(\phi_m - \phi_s)/q$. If an external bias voltage V is applied to the junction (Fig. 8.20), then forward biasing with the semiconductor negative with respect to the metal reduces the energy barrier at the junction by Vq to $(\phi_m - \phi_s - Vq)$ and a current flows through the junction as the barrier over which electrons have to climb from the semiconductor conduction band to fall into the metal conduction band is reduced. If there is reverse biasing with the semiconductor positive with respect to the metal, the energy barrier at the junction is increased by Vq to $(\phi_m + Vq - \phi_s)$ and so electrons from the semiconductor conduction band cannot make it over the barrier into the metal conduction band. This junction, like a p–n junction, is a rectifying junction. A device based on such a junction is called a *Schottky-barrier diode*.

If the metal and the n-type semiconductor have work functions such that $\phi_m < \phi_s$ then the energy bands for such a junction are as shown in Fig. 8.21. Now there is no barrier to electrons from the semiconductor conduction band moving into the metal conduction band. Such a junction is said to be an *ohmic contact*.

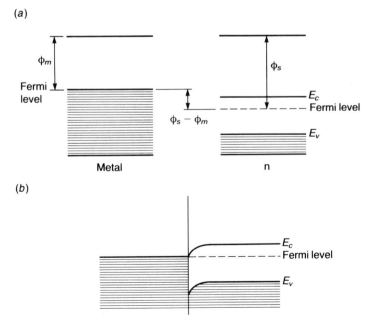

(a)

(b)

Fig. 8.21 Metal and n-type semiconductor junction when $\phi_m < \phi_s$

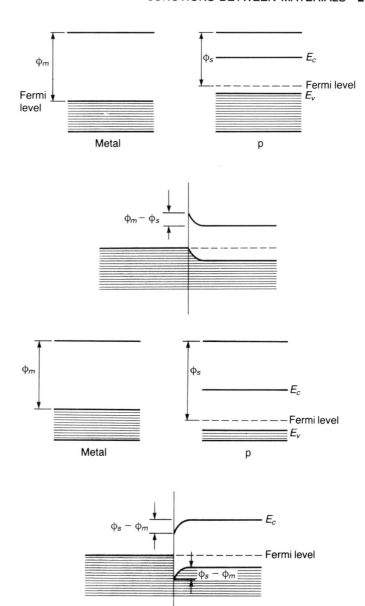

Fig. 8.22 Metal and p-type semiconductor junction when $\phi_m > \phi_s$

Fig. 8.23 Metal and p-type semiconductor junction when $\phi_m < \phi_s$

Figures 8.22 and 8.23 show the energy bands for junctions between a metal and p-type semiconducting materials. In Fig. 8.22 the work function of the metal is greater than that of the semiconductor and the result is an ohmic contact. This is because electrons are easily able to move into the valence band of the semiconductor and consequently holes in the valence band of the semiconductor are able to move directly across into the valence band of the metal. In Fig. 8.23 the work function of the metal is less than that of the semiconductor material and the result is a rectifying junction. This is

Table 8.1 Barrier heights

Semiconductor	Junction with		
	Aluminium	*Gold*	*Copper*
Silicon, n-type	0.7	0.8	0.6
Silicon, p-type	0.4	0.3	0.5
Cadmium sulphide	ohmic	0.8	0.5
Gallium arsenide, n-type	0.8	0.9	0.8
Gallium arsenide, p-type	0.5	0.4	

because holes in the valence band of the semiconductor have to cross the energy barrier $(\phi_s - \phi_m)$. Cuprous oxide is a p-type semiconductor and a rectifier is formed by contact between it and copper. Selenium is also a p-type semiconductor and a rectifier can be formed by contact between it and a suitable alloy of tin.

Table 8.1 shows some typical values of barrier heights for junctions between semiconductors and metals. However, because such junctions are between separate materials, unlike the p–n junction where the junction occurs within a single crystal as a result of differences of doping, the conditions at the surfaces of the materials can markedly affect the barrier heights obtained. The greater the Schottky barrier the better the junction can act as a rectifier. At room temperature a low barrier is easily surmounted by thermally excited charge carriers.

Ohmic contacts

The connection of semiconductor electronic components, such as a p–n junction diode, into an electrical circuit invariably involves metal to semiconductor junctions. For such contacts the junction should not rectify, an ohmic contact being required. Figure 8.24(a) shows the voltage–current characteristic of an ohmic contact, and for comparison in Fig. 8.24(b) the characteristic for a rectifying junction.

As indicated earlier in this chapter, for a metal to p-type semiconductor contact to be ohmic then the work function of the metal has to be more than that of the semiconductor. For a

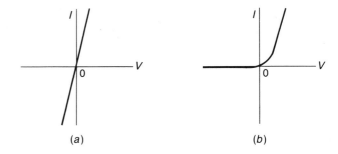

Fig. 8.24 Voltage–current characteristics, (a) ohmic contact, (b) rectifying contact

(a) (b)

metal to n-type semiconductor contact to be ohmic then the work function of the metal has to be less than that of the semiconductor. Thus for p-type silicon aluminium can be used to make the contact because, though it gives a rectifying junction, it gives only a small energy barrier and at room temperature gives essentially an ohmic contact. However, for n-type silicon there are problems in obtaining an ohmic contact, the junction between aluminium and n-type germanium generally being a rectifying junction. This is because the work function of n-type silicon is generally less than that of metals. This can be overcome by heavy doping of the silicon at the interface, to the extent of making the semiconductor degenerate, i.e., like a metal rather than semiconductor (see Ch. 7). With contact between a metal and such a degenerate semiconductor the depletion band is so narrow that electrons are able to get through the energy barrier, the term used being to 'tunnel' through it. The junction is then ohmic.

Photodiode

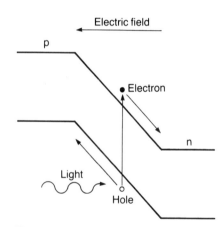

Fig. 8.25   Photodiode

A *photodiode* is just a p–n junction in a transparent encapsulation. The junction is reverse biased (Fig. 8.25). Under such conditions the current I_0 across the junction is very low. However, if light of a suitable frequency is incident on the junction then it is able to excite electrons from the valence band to the conduction band. The electric field across the junction is in such a direction as to sweep the electrons into the n region and the holes into the p region. The charge carriers produced by the light thus give rise to a current which adds directly to the current I_0, the total current then being proportional to the intensity of illumination.

A photon can only cause an electron to jump from the valence band to the conduction band if it has sufficient energy (see earlier discussion of the photoconductive cell). The energy of a photon is given by $E = hf$, where h is Planck's constant and f the frequency. Thus if E_g is the energy gap between the bands then the minimum frequency is given by

$$E_g = hf_{min}$$

Since the frequency is related to the wavelength λ by $c = f\lambda$, where c is the speed of light, then the minimum frequency means a maximum wavelength λ_{max}, where

$$E_g = \frac{hc}{\lambda_{max}} \qquad [30]$$

The energy gap for silicon is such as to mean that a silicon photodiode is not sensitive to light with wavelength over about 1.1 μm. Visible light has wavelengths in the region 0.4 to

0.7 μm. To detect light in this region the photodiode used is a metal–semiconductor junction, e.g., gold–silicon.

Example 12

What will be the maximum wavelength to which a gallium arsenide photodiode will respond if it has an energy gap of 1.4 eV?
Planck's constant = 6.6×10^{-34} J s
Speed of light = 3.0×10^{8} m/s

Answer

Using equation [30]

$$E_g = \frac{hc}{\lambda_{max}}$$

Thus

$$\lambda_{max} = \frac{6.6 \times 10^{-34} \times 3.0 \times 10^{8}}{1.4 \times 1.6 \times 10^{-16}} = 8.8 \times 10^{-7} \text{ m}$$

Example 13

What is the maximum wavelength to which a gold–silicon photodiode will respond if the Schottky barrier height is 0.8 eV?
Planck's constant = 6.6×10^{-34} J s
Speed of light = 3.0×10^{8} m/s

Answer

The Schottky barrier is the energy E_{Sh} through which an electron has to be moved as a result of receiving energy from a photon. Thus, in a similar way to equation [30]

$$E_{Sh} = \frac{hc}{\lambda_{max}}$$

Thus

$$\lambda_{max} = \frac{6.6 \times 10^{-34} \times 3.0 \times 10^{8}}{0.8 \times 1.6 \times 10^{-16}} = 1.5 \times 10^{-6} \text{ m}$$

Light-emitting diode

A forward biased p–n junction has a current flowing through its junction as a result of holes from the p side of the junction moving into the n side and electrons from the n side moving into the p side. A hole in the n-type material does not last very long and is soon occupied by an electron. Similarly an electron in the p-type material soon recombines with a hole. This recombination (Fig. 8.26) involves an electron falling from the conduction band to the valence band or from an electron trap to a hole trap in a material so doped (see earlier in this chapter). In doing this the electron has to lose energy and it can do this by emitting a photon (since both momentum and

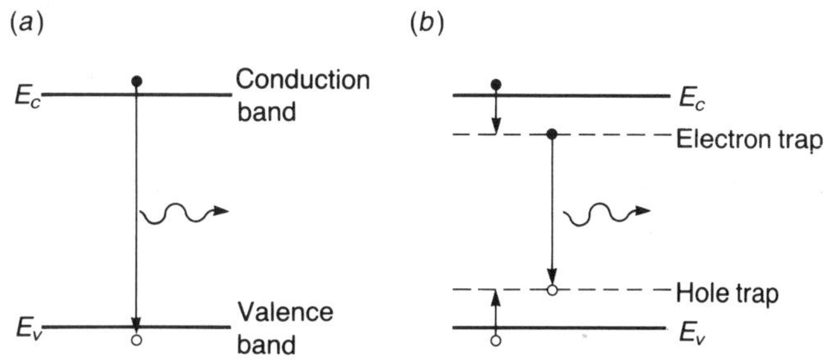

(a)

(b)

Fig. 8.26 Light-emitting diodes, photon emission from (a) conduction band to valence band transition, (b) electron trap to hole trap transition

energy have to be conserved, in some cases phonons can also be emitted). Such light-emitting diodes are generally just referred to by their initials, i.e. LEDs.

Where all the energy is used for the emission of a photon the frequency f of the photon will be given by

$$E = hf$$

where E is the energy lost by the falling electron and h is Planck's constant. Since $c = f\lambda$, the wavelength λ is given by

$$E = \frac{hc}{\lambda} \qquad [31]$$

where c is the speed of light.

Gallium arsenide, gallium phosphide and alloys of gallium arsenide with gallium phosphide are widely used for light-emitting diodes. The gallium arsenide LED has a junction formed by diffusing the acceptor zinc into n-type gallium arsenide and emits light of wavelength about 0.9 µm, i.e., in the infra-red. Such an LED is widely used in television remote controls to communicate by means of infra-red radiation between the control unit and the television set. The gallium phosphide LED when doped with nitrogen emits light of wavelength about 0.57 µm (green light) with light doping and 0.59 µm (yellow light) with heavy doping. A 60% gallium arsenide–40% gallium phosphide alloy is used for a red LED, a 35% gallium arsenide–65% gallium phosphide alloy for an orange LED and a 15% gallium arsenide–85% gallium phosphide for a yellow LED.

Example 14

What is the energy difference between the levels between which an electron in an LED falls if it emits light of wavelength 0.57 µm? Planck's constant $= 6.6 \times 10^{-34}$ J s
Speed of light $= 3.0 \times 10^8$ m/s

Answer

Using equation [31]

$$E = \frac{hc}{\lambda} = \frac{6.6 \times 10^{-34} \times 3.0 \times 10^{8}}{0.57 \times 10^{-6}} = 3.5 \times 10^{-19} \text{ J} = 2.2 \text{ eV}$$

Semiconductor injection laser

The *semiconductor injection laser* is a development of the light-emitting diode (LED). The term laser stands for *l*ight *a*mplification by *s*timulated *e*mission of *r*adiation. When a photon interacts with an electron there are two possible outcomes (Fig. 8.27). The photon can be absorbed and its energy used to excite the electron to a higher energy level or if the electron is already at this higher energy level the photon can cause the electron to drop down to a lower energy level and emit a second photon, the second photon being identical in all respects to the incident photon. This is called *stimulated emission* and results in an amplification of the number of photons. For stimulated emission to occur the p–n junction must already have a large population of electrons in the conduction band just waiting to recombine with holes in the valence band, more than the population of electrons in the valence band waiting to be moved up into the conduction band. This can be achieved by the use of high doping densities and a high forward bias. These conditions lead to large numbers of electrons in the valence band being 'injected' into the conduction band.

(a)   (b)

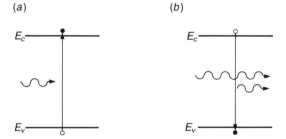

Fig. 8.27 Interaction between a photon and an electron in a semiconductor, (a) excitation, (b) stimulated emission

Figure 8.28(a) shows the basic form of a semiconductor injection laser. The two end faces of the laser are polished so that they internally reflect back into the semiconductor about 30% of the photons, the sides of the laser are roughened so that no such reflection occurs and all the photons moving back and forth through the laser are in the same line through the p–n junction. A spontaneous first emission of a photon by an electron falling from the conduction band to the valence band leads to a photon which instead of being absorbed or radiated out of the junction is trapped between the reflecting end faces and so is able to stimulate the emission of another identical photon. These two photons can then lead to further stimulated emission and yet more photons. The result is light amplification.

(a)

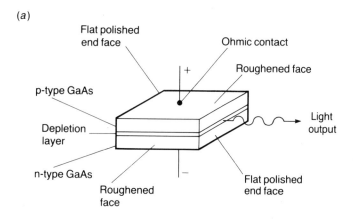

(b)

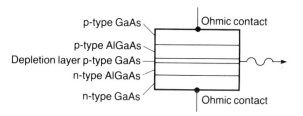

Fig. 8.28 (a) Basic semiconductor injection laser, (b) Multilayered semiconductor injection laser

The light intensity reaches an equilibrium value when the light losses from the junction balance the light amplification.

A gallium arsenide injection laser emits radiation at a wavelength of about 0.9 μm, this being the result of electron movements between the conduction band and valence band with an energy gap of about 1.4 eV. A problem with this form of laser is the very high current densities occurring as a result of the high forward bias. This problem can be reduced by the use of multiple layers, rather than just one layer of p-type semiconductor and one layer of n-type semiconductor. Figure 8.28(b) shows such a laser involving layers of gallium arsenide and aluminium gallium arsenide. Such a laser is widely used in compact-disc players.

Junction field-effect transistor

The basic form of the *junction field-effect transistor* (JFET) is shown in Fig. 8.29. It is essentially a strip of semiconducting material between two ohmic contacts called the source and the drain. The current path between the source and drain is called the channel. The current is restricted to this path since it is bounded on both sides by the depletion layers of p–n junctions. Within the depletion layer there are virtually no

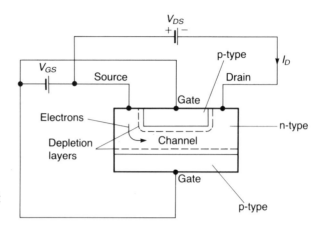

Fig. 8.29 Junction field-effect transistor

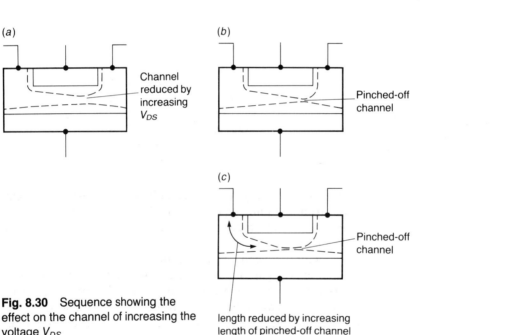

Fig. 8.30 Sequence showing the effect on the channel of increasing the voltage V_{DS}

free charge carriers and so it has a high resistivity. The depletion layers thus isolate the channel. The width of each depletion layer can be increased by increasing the reverse bias across it, i.e., by increasing the voltage V_{DS} with the gate voltage V_{GS} being maintained constant. Such a change reduces the width of the channel and thus changes the resistance of the path between the source and drain. Indeed if the depletion layers are increased sufficiently the channel can be virtually pinched off (Fig. 8.30). It is not completely cut off but increasing the voltage V_{DS} above the pinch-off value just reduces the length of the channel outside the pinched-off

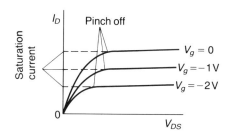

Fig. 8.31 Drain current plotted against drain–source voltage

region. A graph of the current through the channel I_D plotted against V_{DS} is thus of the form shown in Fig. 8.31, showing that as V_{DS} increases so the resistance increases until it becomes very high at the pinch-off value. There is only a slight change in the resistance as the voltage is further increased. The value of the virtually constant current beyond the pinch-off point is called the saturation current. A family of such graphs are produced for different values of the gate voltage V_G. Making V_G more negative reduces the voltage V_{DS} required for pinch off. A consequence of this is that the saturation value of the current I_D depends on the value of the gate voltage. The gate voltage can thus be used to control the current in the source–drain circuit.

Metal–oxide–silicon field-effect transistor

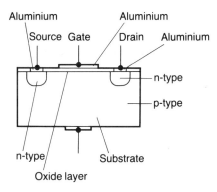

Fig. 8.32 The metal–oxide–silicon field-effect transistor

The *metal–oxide–silicon field-effect transistor* is referred to as a MOSFET, or MOS transistor (MOST), or insulated gate FET (IGFET). One form of such a transistor is shown in Fig. 8.32, this form being called the n-channel form. It consists of a lightly doped p-type substrate into which two heavily doped n-type regions are diffused to form two p–n junctions. Connections to these two regions via aluminium ohmic contacts provide the source and drain. Separating the source and drain is a region of p-type material coated with a thin surface layer of silicon oxide. The oxide is an electrical insulator with a relative permittivity of about 3.9. The upper surface of the oxide is covered with a thin layer of aluminium to which the gate contact is made. The aluminium–oxide–p substrate forms a capacitor with the oxide as the dielectric.

With no gate-to-substrate bias the position is of two p–n junctions separated by the lightly doped p substrate, which because of the light doping and hence low numbers of charge carriers has a high resistance. There is, in this situation, no low-resistance channel between the source and the drain. The substrate is normally held at the same potential as the source, hence no bias across the source-to-substrate p–n junction. The drain is at a positive potential with respect to both the source and the substrate and hence the drain-to-substrate p–n junction is reverse biased. Thus in this situation (Fig. 8.33(a)) there is virtually no current between the source and drain.

If a positive voltage is applied to the gate then, as a consequence of the potential difference across the metal–oxide–silicon capacitor, an electric field is produced in the oxide dielectric and the metal becomes positively charged and the silicon side of the capacitor negatively charged. This negative charge means that electrons have been attracted to the region and holes repelled. If the gate voltage is high enough the low number of holes in the p substrate at this region can become

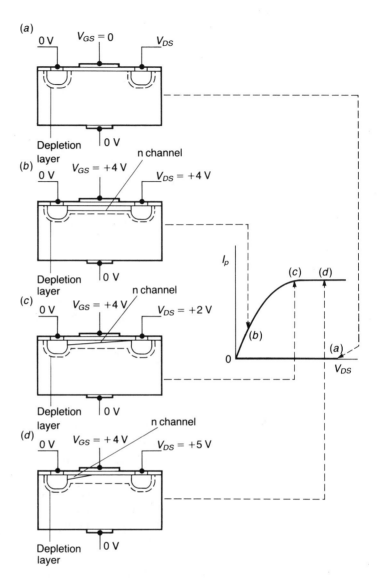

Fig. 8.33 The behaviour of a MOSFET at different drain–source voltages

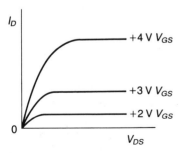

Fig. 8.34 The current–voltage characteristic

swamped by the electrons and become changed into an n-type material (Fig. 8.33 (*b*)). The layer is said to have become *inverted*. As a consequence there is then a channel of n-type material linking the source and drain n-type regions. Within all this n-type material there are free electrons. A voltage between the drain and the source then results in a current.

If the drain to source voltage V_{DS} is increased the current through the channel increases but also the voltage between the drain and gate decreases. As a consequence of this the potential difference across the oxide capacitor decreases and so the charge on the 'plates' of the capacitor decreases. As a consequence, not only does the density of electrons in the channel decrease and thus its resistance increase, but because there is a potential drop along the length of the channel the drain end of the oxide capacitor may not have sufficient electrons to be inverted to n-type material (Fig. 8.33 (*c*)). Thus the resistance increases quite rapidly as V_{DS} is increased. At a high enough value of V_{DS} the resistance of the channel may become so high that there is no significant increase in current with an increase in V_{DS} (Fig. 8.33(*d*)).

Figure 8.34 shows the type of relationship that occurs between the current I_D through the channel, i.e., between source and drain, when the potential difference between the drain and source V_{DS} is increased. A change in the positive voltage V_{GS} applied to the gate results in a change in the potential difference across the oxide capacitor and hence the density of conduction electrons in the channel. Thus there are different current–V_{DS} graphs for different gate voltages.

The above discussion is of a MOSFET in which a n-type channel is produced in a p substrate. A similar situation can be produced for a p-type channel in an n substrate.

Bipolar junction transistor

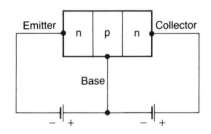

Fig. 8.35 The n–p–n transistor

The *bipolar junction transistor* consists of two back-to-back p–n junctions in a single piece of semiconductor material. There are two possible forms, the p–n–p and the n–p–n. Figure 8.35 is a diagrammatic representation of the n–p–n transistor and shows how the two p–n junctions are normally biased. The emitter to base p–n junction is forward biased and the collector to base junction reverse biased. Because of this biasing we can think of the forward-biased junction as a source, an emitter, of electrons which are injected into the reverse-biased junction.

Figure 8.36 shows the energy band picture for the n–p–n transistor before biasing and after biasing. The effect of the biasing has been to reduce the barrier height between the emitter and the base and increase it between the collector and the base. This reduction in barrier height between emitter and base allows electrons in the emitter to move into the base. The emitter is more heavily doped than the base and so ensures that the electron current from the emitter to base is considerably larger than any hole current from base to emitter. These electrons diffuse across the base, together with minority electrons already in the base, and become swept through into the collector. Thus an electron current through the low-resistance forward-biased emitter–base junction has become a current through the high-resistance reverse-biased collector–base junction. The origin of the term transistor is 'transfer-resistor' and refers to this transfer of a signal from a low resistance to a high resistance.

If a fraction α of the emitter current I_E reaches the collector

(a)

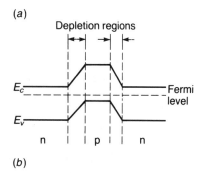

(b)

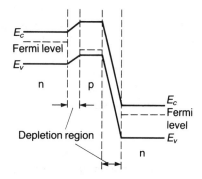

Fig. 8.36 Energy bands for n–p–n transistor, (a) before bias, (b) biased

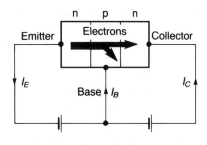

Fig. 8.37 Currents with a biased n–p–n transistor

then the collector current due to this is αI_E. The fraction α, generally termed the *common base current gain*, has typically a value between about 0.95 and 0.995. The reasons for α being less than 1 is that the biasing of the emitter–base junction results in the movement of some holes from the base to the emitter, also some of the electrons diffusing across the base will recombine with holes present as majority carriers. The collector current I_C must also include the normal reverse bias current I_{CB0} that occurs with the reverse-biased collector–base junction. Thus

$$I_C = \alpha I_E + I_{CB0} \tag{32}$$

Since I_{CB0} is very small, it is generally neglected and the equation becomes

$$\alpha \approx \frac{I_C}{I_E} \tag{33}$$

The term *emitter efficiency* γ is used for the ratio of the current due to electrons I_{En} and the total current, i.e. the sum of the current due to electrons I_{En} and due to holes I_{Ep}, crossing the emitter–base junction.

$$\gamma = \frac{I_{En}}{I_{En} + I_{Ep}} \tag{34}$$

The heavy doping of the emitter in relation to the lighter doping of the base ensures that the hole current remains fairly small in relation to the electron current and thus γ is close to 1.

If recombination is negligible it is a reasonable approximation to equate the electron current through the emitter–base junction with the collector current, i.e., $I_{En} \approx I_C$. Thus since the total emitter current is I_E then the emitter efficiency approximates to

$$\gamma \approx \frac{I_C}{I_E} \approx \alpha$$

The transistor is just essentially a junction at which a collector current and a base current flow in and an emitter current out (Fig. 8.37). Note that by convention the current directions are in the opposite direction to the directions of electron movement. Thus

$$I_C = I_E - I_B \tag{35}$$

Thus using equation [33], $I_C = \alpha I_E$, then

$$\alpha I_E = I_E - I_B$$

$$I_B = I_E(1 - \alpha)$$

Thus since α is very close to 1, the base current is very small.

The above discussion all refers to an n–p–n transistor. A similar situation, and similar definitions, apply to a p–n–p transistor. The main difference is that the current through the transistor is mainly carried by holes rather than electrons.

Example 15

A transistor has a common base current gain of 0.99. What percentage of the current through the emitter–base junction will be holes?

Answer

Since the emitter efficiency can reasonably be assumed to be the same as the common base current gain, then equation [34] indicates

$$\gamma = \frac{I_{En}}{I_{En} + I_{Ep}} = 0.99$$

Thus

$$\frac{I_{En} + I_{Ep}}{I_{En}} = \frac{1}{0.99}$$

$$\frac{I_{Ep}}{I_{En}} = \frac{1}{0.99} - 1 = \frac{0.01}{0.99} \approx 0.01$$

Thus 1% of the current can be expected to be holes.

Problems

1 Indium antimonide when used as a photoconductive resistor responds to a maximum wavelength of 5.5 μm. What is the size of the energy gap between the valence and conduction bands? Planck's constant $= 6.6 \times 10^{-34}$ J s, speed of light $= 3.0 \times 10^8$ m/s.

2 Why are photoconductive cells used for the detection of wavelengths in the infra-red generally cooled?

3 What is the hole diffusion current density in a semiconductor if the hole density changes from 3.0×10^{10} holes/m^3 to 2.0×10^9 holes/m^3 in a distance of 10 μm? $D_p = 1.2 \times 10^{-3}$ m^2/s and the charge carried by a hole is 1.6×10^{-19} C.

4 What are the diffusion coefficients for electrons and holes in germanium at 300 K where the electron mobility is 0.39 m^2 V^{-1} s^{-1} and the hole mobility 0.19 m^2 V^{-1} s^{-1}? Boltzmann's constant $= 1.38 \times 10^{-23}$ J/K, charge carried by a hole or electron $= 1.6 \times 10^{-19}$ C.

5 What are the diffusion lengths for excess electrons and holes in silicon if they have a lifetime of 1.0×10^{-6} s? For silicon $D_n = 3.9 \times 10^{-3}$ m^2/s, $D_p = 1.2 \times 10^{-3}$ m^2/s.

6 What is the contact potential difference at a junction between zinc and copper if zinc has a work function of 4.1 eV and copper 4.5 eV?

7 What is the equilibrium potential difference across a germanium

p–n junction at 300 K if each side of the junction has a doping density of 10^{22} dopants per cubic metre and the intrinsic concentration for germanium is 2.4×10^{19} /m^3? Boltzmann's constant $= 1.38 \times 10^{-23}$ J/K, charge carried by a hole or electron $= 1.6 \times 10^{-19}$ C.

8 A silicon p–n junction diode at 300 K has a saturation current of 1.0×10^{-13} A. What will be the current through the diode when it is forward biased by a potential difference of 0.6 V? Boltzmann's constant $= 1.38 \times 10^{-23}$ J/K, charge carried by a hole or electron $= 1.6 \times 10^{-19}$ C.

9 An abrupt silicon p–n junction has an acceptor density of 2.0×10^{23} /m^3 and a donor density of 1.0×10^{22} /m^3. What is the width of the depletion region at 300 K when the junction has (a) a reverse bias of 10 V, (b) a forward bias of 0.50 V?
The relative permittivity of silicon is 12.
Permittivity of free space $= 8.9 \times 10^{-12}$ F/m.
Boltzmann's constant $= 1.38 \times 10^{-23}$ J/K.
Intrinsic concentration for silicon at 300 K $= 1.4 \times 10^{16}$ /m^3.
Charge carried by an electron or hole $= 1.6 \times 10^{-19}$ C.

10 What is the effect on the width of the depletion layer in a p–n junction of (a) biasing and (b) a change in doping density?

11 Sketch the energy bands, indicating the position of the Fermi level, for a p–n junction in (a) equilibrium, (b) reverse biased, and (c) forward biased. Explain the significance of the differences produced by the biasing.

12 What is the height of the energy hill between the bottom levels of the conduction bands in the p and n materials of a p–n junction at (a) equilibrium, and (b) with a reverse bias of 4.0 V, if the contact potential between the two is 0.30 V?

13 Explain why the junction between cuprous oxide and copper can be used as a rectifier.

14 What will be the maximum wavelength to which a silicon photodiode will respond if it has an energy gap of 1.1 eV? Planck's constant $= 6.6 \times 10^{-34}$ J s, speed of light $= 3.0 \times 10^8$ m/s.

15 An LED is required for the emission of light of wavelength 0.6 μm. What will be the difference in energy between the levels between which electrons must fall to give such an emission? Planck's constant $= 6.6 \times 10^{-34}$ J s, speed of light $= 3.0 \times 10^8$ m/s.

16 Explain the principles involved in the operation of a semiconductor injection laser.

17 Explain how the channel width in a JFET can be changed and its consequential effect on the current through it.

18 Explain what is meant by and the significance of inversion in a MOSFET.

19 Sketch the energy band diagrams for a p–n–p transistor (a) under no bias conditions and (b) with the emitter–base junction forward biased and the collector–base junction reverse biased.

Answers to problems

Chapter 1

1. (*a*) One carbon, two oxygen. (*b*) One carbon, four hydrogen. (*c*) Two nitrogen
2. Z = number of protons, M = number or protons + number of neutrons, isotopes have same numbers of protons but different numbers of neutrons
3. (*a*) 20 electrons, 20 protons, 20 neutrons, (*b*) 26 electrons, 26 protons, 30 neutrons, (*c*) 50 electrons, 50 protons, 70 neutrons
4. Filled shells give stability
5. n: 1, 2, 3,. . .; l: (n − 1); m_l: 0, −l, +l; m_s: +½, −½. See text
6. Two
7. See, for example, Figs 1.5 and 1.6
8. (*a*) 3, (*b*) 3, (*c*) 2
9. (*a*) Insulator, (*b*) good conductor
10. (*a*) Metallic, (*b*) covalent, (*c*) ionic, (*d*) ionic, (*e*) metallic, (*f*) Van der Waals
11. Chromium and molybdenum
12. As graphite, one valency electron is not covalently bonded

Chapter 2

1. See Fig. 2.5
2. See text
3. 1.2×10^{-4} C
4. (*a*) 1.6×10^{-5} C, (*b*) 8 V, 4 V
5. 14 μF
6. See text
7. 11 V/m radially
8. Separation, area, dielectric
9. 5.0×10^{-10} F
10. (*a*) 6.5×10^{-10} C, (*b*) 9.8×10^{-6} C/m^2, (*c*) 1.6×10^4 V
11. 2.5×10^4 V/m

12. 5.0×10^4 V/m, 1.1×10^{-6} C/m^2
13. See Fig. 2.13
14. (a) 6.0×10^{-10} F, (b) 7.2×10^3 V/m, 3.6×10^3 V/m
15. 1.7×10^{-9} F
16. See text
17. 1.0×10^{-4} J
18. 1.4 V
19. (a) 2.95×10^{-6} C/m^2, (b) 0.49 J/m^3

Chapter 3

1. See text
2. 1.7×10^{-8} C/m^2
3. Frictional effects and current leakage
4. See text
5. 0.22 W
6. 2.3×10^8 Ω
7. 100 V
8. 4.6×10^{-11} F
9. High dielectric strength, high surface/volume resistivity, low loss factor, high permittivity
10. See text
11. 1.9 kV
12. (a) 0.3 V, (b) 7.0×10^{-8} C/m^2

Chapter 4

1. 8.6 V
2. 2400 A
3. 25 000 A/m
4. (a) 2.39×10^6 A/Wb, (b) 2.09×10^{-4} Wb
5. (a) 1714 A/m, (b) 2.15 mT, (c) 9.69×10^{-7} Wb
6. 796 A
7. (a) 2.56×10^6 A/Wb, (b) 5.12 A
8. (a) 2.26×10^6 A/Wb, (b) 4.55×10^6, (c) 6.4 A
9. 2.78 A
10. 5.39×10^{-4} Wb
11. (a) Central limb twice outer limb, (b) the same, (c) 0.42 T
12. 0.96 A
13. 0.98 A
14. 60 V
15. 36 mH
16. 6.8×10^{-4} H
17. 2.77×10^{-7} H/m
18. 3.27×10^{-7} H/m
19. 1.8 μH/m
20. 4 V
21. 0.35
22. (a) 20, (b) 3 A

23. (*a*) 1.8×10^{-3} Wb, (*b*) 400 V
24. See Fig. 4.21 and Fig. 4.22
25. See text
26. See text and Fig. 4.25 and Fig. 4.26
27. 2.48 Ω
28. 1.4
29. 95.5%
30. See text
31. 3 W
32. 6.3 W
33. 20 W, 100 W
34. (*a*) 1.2 N/m, (*b*) zero
35. 1.5 mJ
36. 19 N

Chapter 5

1. (*a*) 3.5×10^4 A/m, (*b*) 0.44 T
2. Diamagnetic: $\mu_r < 1$, negative susceptibility; paramagnetic $\mu_r > 1$, susceptibility small and positive; ferromagnetic $\mu_r \gg 1$, susceptibility large and positive
3. Iron has unpaired, coupled electrons and is ferromagnetic, copper does not and is diamagnetic
4. Ferromagnetic: dipoles in neighbouring atoms aligned in same direction, antiferromagnetic dipoles aligned in opposite directions
5. (*a*) 1.0 T, (*b*) 3×10^5 A/m, (*c*) 1.2 T, (*d*) approx. 17×10^5 J/m, (*e*) approx. 1.4×10^5 T A/m
6. Ferrites have much higher resistivities than ferromagnetics due to ionic rather than metallic bonding in the solid
7. 0.020 m^3
8. (*a*) and (*b*) high permeability, low remanence, low coercive field, low hysteresis loss. For (*b*) a Fe–78% Ni alloy or a ferrite are often used. (*c*) Square loop, low remanence, low coercive field. A ferrite based on magnetite is often used

Chapter 6

1. See text in Chapter 6
2. (*a*) 4.6×10^{14} Hz, (*b*) 651 nm, (*c*) 9.2×10^{20} J, (*d*) 4.5×10^5 m/s
3. 1.0×10^7 m/s
4. 5.9×10^6 m/s
5. 8.0×10^{-15} N
6. 15.6 mm
7. 0.84 m
8. 3.4×10^{-5} T
9. Helical

10. 15.8 mT
11. 14 mm
12. 0.05 T

Chapter 7

1. See Fig. 7.3
2. SiO_2 an insulator, Si a semiconductor
3. (a) p-type, (b) n-type, (c) p-type, (d) n-type, (e) n-type, (f) n-type
4. $n = 2.0 \times 10^{10}$ /m³, $p = 10^{22}$ /m³
5. 1.5×10^{-4} m/s
6. Increased by factor of 1.4
7. (a) 4.0×10^{-3} m² V⁻¹ s⁻¹, (b) 3.6×10^{-2} V/m, (c) 1.4×10^{-4} m/s
8. 0.45 Ω m
9. 1.6×10^{-4} Ω m
10. See text and Figs 7.18, 7.19, 7.20
11. 0.35 eV above intrinsic level
12. See Fig. 7.21 and 7.24 and text
13. See text and example 12
14. 8.5×10^{28} /m³
15. (a) 3.2×10^{21} /m³, (b) 1.9×10^{-3} m³/C

Chapter 8

1. 0.23 eV
2. See text, prevent thermally excited electrons swamping light-excited electrons
3. 5.4×10^{-7} A/m²
4. 1.0×10^{-2} m²/s, 4.9×10^{-3} m²/s
5. 6.2×10^{-5} m, 3.5×10^{-5} m
6. 0.4 V
7. − 0.43 V
8. 1.2 mA
9. (a) 1.2×10^{-6} m, (b) 2.0×10^{-7} m
10. (a) Reverse bias increases width, forward bias decreases it, (b) increase in doping density decreases width
11. See Fig. 8.18 and text
12. (a) 0.30 eV, (b) 4.30 eV
13. See text, $\phi_m < \phi_s$
14. 1.2 μm
15. 2.1 eV
16. See text
17. See text
18. See text
19. (a) A valley, (b) the right-hand side of the valley being elevated and the left lowered

Index